Lipoprotein Analysi

D1765712

UNIVERSITIES AT MEDWAY
DRILL HALL LIBRARY
ORDINARY LOAN

This book must be returned or renewed by the last date stamped below, and may be recalled earlier if needed by other readers.

Fines will be charged as soon as it becomes overdue.

UNIVERSITIES
at MEDWAY

D4187-10

The Practical Approach Series

SERIES EDITORS

D. RICKWOOD
Department of Biology, University of Essex
Wivenhoe Park, Colchester, Essex CO4 3SQ, UK

B. D. HAMES
Department of Biochemistry and Molecular Biology, University of Leeds
Leeds LS2 9JT, UK

Affinity Chromatography
Anaerobic Microbiology
Animal Cell Culture
  (2nd Edition)
Animal Virus Pathogenesis
Antibodies I and II
Biochemical Toxicology
Biological Membranes
Biomechanics—Materials
Biomechanics—Structures and
  Systems
Biosensors
Carbohydrate Analysis
Cell Growth and Division
Cellular Calcium
Cellular Neurobiology
Centrifugation (2nd Edition)
Clinical Immunology
Computers in Microbiology
Crystallization of Proteins and
  Nucleic Acids
Cytokines
The Cytoskeleton

Diagnostic Molecular Pathology
  I and II
Directed Mutagenesis
DNA Cloning I, II, and III
Drosophila
Electron Microscopy in
  Biology
Electron Microscopy in
  Molecular Biology
Enzyme Assays
Essential Molecular Biology
  I and II
Fermentation
Flow Cytometry
Gel Electrophoresis of Nucleic
  Acids (2nd Edition)
Gel Electrophoresis of Proteins
  (2nd Edition)
Genome Analysis
HPLC of Macromolecules
HPLC of Small Molecules
Human Cytogenetics I and II
  (2nd Edition)

Lipoprotein
Analysis
A Practical Approach

Edited by
CAROLYN A. CONVERSE
Department of Pharmaceutical Sciences, University of Strathclyde,
204 George Street, Glasgow, G1 1XW, UK

and

E. ROY SKINNER
Department of Molecular and Cell Biology, University of Aberdeen,
Marischal College, Aberdeen AB9 1AS, UK

IRL PRESS
——at——
OXFORD UNIVERSITY PRESS
Oxford New York Tokyo

Oxford University Press, Walton Street, Oxford OX2 6DP

Oxford is a trade mark of Oxford University Press

Published in the United States
by Oxford University Press, New York

© *Oxford University Press, 1992*

All rights reserved. No part of this publication may be reproduced,
stored in a retrieval system, or transmitted, in any form or by any means,
electronic, mechanical, photocopying, recording, or otherwise, without
the prior permission of Oxford University Press

This book is sold subject to the condition that it shall not, by way
of trade or otherwise, be lent, re-sold, hired out or otherwise circulated
without the publisher's prior consent in any form of binding or cover
other than that in which it is published and without a similar condition
including this condition being imposed on the subsequent purchaser

Users of books in the Practical Approach series are advised that
prudent laboratory safety procedures should be followed at all times.
Oxford University Press make no representation, express or implied,
in respect of the accuracy of the material set forth in books in this series
and cannot accept any legal responsibility or liability for any errors
or omissions that may be made.

British Library Cataloguing in Publication Data
A cataloguing record for this title is
available from the British Library

Library of Congress Cataloging in Publication Data
Lipoprotein analysis: a practical approach/edited by Carolyn A.
Converse and E. Roy Skinner.
p. cm.—(Practical approach series)
Includes bibliographical references and index.
1. Lipoproteins—Analysis. 2. Lipoproteins—Separation.
3. Lipoproteins—Research—Methodology. I. Converse, Carolyn A.
II. Skinner, E. Roy. III. Series.
[DNLM: 1. Lipoproteins—analysis. 2. Lipoproteins—metabolism.
QU 85 L7654]
QP552.L5L543 1992 612.3'97—dc20 91–35365
ISBN 0–19–963192–1 (h/b)
ISBN 0–19–963231–6 (p/b)

Typeset by Cambrian Typesetters, Frimley, Surrey
Printed in Great Britain by
Information Press Ltd, Eynsham, Oxford

Preface

THE increasing awareness over recent years that defects in lipoprotein metabolism, particularly in the action of lipoprotein receptors, are associated with the development of coronary heart disease, the major cause of death in most western countries, has led to an enormous proliferation in lipoprotein research. Investigations have been made on a broad front, ranging from intensive clinical investigations to detailed studies on apolipoproteins that confer specificity to the lipoprotein particles, the nature and metabolism of the latter, and the action of enzymes and receptors that control the course of lipoprotein metabolism. Although these investigations have resulted in a vast literature concerning the numerous aspects of the subject, many unanswered questions still remain to be solved before we can properly understand the role of lipoproteins in maintaining the normal distribution of lipids throughout the body and the transport of associated lipophilic compounds, as well as in the development of atherosclerosis. The aim of this book is to provide details of the wide variety of techniques that are required for such studies.

Methods required for studies on the molecular biology of lipoproteins have not been included. Although this is a highly important and rapidly developing area of lipoprotein research, we feel that the subject is too vast for inclusion in a book of this size without detriment to other areas; companion volumes in this series provide an excellent account of the techniques used in such studies.

We are indebted to our co-authors who have made this book a useful source of information which we hope will be of value to those facing the many fascinating challenges that lie ahead in lipoprotein research.

Glasgow and Aberdeen   C. A. C.
January 1992   E. R. S.

Contents

3. Immunological methods for studying and quantifying lipoproteins and apolipoproteins   61

Ross W. Milne, Philip K. Weech, and Yves L. Marcel

6. Lipoprotein–receptor interactions   145

Kay S. Arnold, Thomas L. Innerarity, Robert E. Pitas, and Robert W. Mahley

7a. Assay of lipoprotein lipase and hepatic lipase   169

Gunilla Bengtsson-Olivecrona and Thomas Olivecrona

7b. Cholesterol esterifying enzymes—lecithin: cholesterol acyltransferase (LCAT) and acylcoenzyme A:cholesterol acyltransferase (ACAT)   187

Michael P. T. Gillett and James S. Owen

7c. Assay of 3-hydroxy-3-methylglutaryl coenzyme A (HMG-CoA) reductase   203

Peter A. Wilce and Paulus A. Kroon

8. Analysis of tissue lipoproteins 215

Robert E. Pitas and Robert W. Mahley

Contents

Contributors

KAY S. ARNOLD
Gladstone Foundation Laboratories for Cardiovascular Disease, Cardiovascular Research Institute, University of California at San Francisco, PO Box 40608, San Francisco, California 94140, USA

GUNILLA BENGTSSON-OLIVECRONA
Department of Medical Biochemistry and Biophysics, University of Umeå, S-901 87 Umeå, Sweden

HANS DIEPLINGER
Institute of Medical Biology and Human Genetics, University of Innsbruck, Schöpfstrasse 41, A-6020 Innsbruck, Austria

PAUL N. DURRINGTON
University of Manchester Department of Medicine, Manchester Royal Infirmary, Oxford Road, Manchester M13 9WL, UK

ALLAN GAW
Institute of Biochemistry, Royal Infirmary, Glasgow G4 0SF, UK

MICHAEL P. T. GILLETT
Academic Department of Medicine, Royal Free Hospital School of Medicine, Rowland Hill Street, London NW3 2PF, UK

THOMAS L. INNERARITY
Gladstone Foundation Laboratories for Cardiovascular Disease, Cardiovascular Research Institute, University of California at San Francisco, PO Box 40608, San Francisco, California 94140, USA

PAULUS A. KROON
Clinical Research Centre, Royal Brisbane Hospital Foundation and Department of Biochemistry, University of Queensland, Queensland, 4072, Australia

MICHAEL I. MACKNESS
University of Manchester Department of Medicine, Manchester Royal Infirmary, Oxford Road, Manchester M13 9WL, UK

ROBERT W. MAHLEY
Gladstone Foundation Laboratories for Cardiovascular Disease, Cardiovascular Research Institute, University of California at San Francisco, PO Box 40608, San Francisco, California 94140, USA

Contributors

YVES L. MARCEL
Laboratory of Lipoprotein Metabolism, Institut de Recherches Cliniques de Montréal, 110 Pine Avenue West, Montréal, Québec, Canada H2W 1R7

HANS-JÜRGEN MENZEL
Institute of Medical Biology and Human Genetics, University of Innsbruck, Schöpfstrasse 41, A-6020 Innsbruck, Austria

ROSS W. MILNE
Laboratory of Lipoprotein Metabolism, Institut de Recherches Cliniques de Montréal; 110 Pine Avenue West, Montréal, Québec, Canada H2W 1R7

THOMAS OLIVECRONA
Department of Medical Biochemistry and Biophysics, University of Umeå, S-901 87 Umeå, Sweden

JAMES S. OWEN
Academic Department of Medicine, Royal Free Hospital School of Medicine, Rowland Hill Street, London NW3 2PF, UK

CHRISTOPHER J. PACKARD
Institute of Biochemistry, Royal Infirmary, Glasgow G4 0SF, UK

ROBERT E. PITAS
Gladstone Foundation Laboratories for Cardiovascular Disease, Cardiovascular Research Institute, University of California at San Francisco, PO Box 40608, San Francisco, California 94140, USA

JAMES SHEPHERD
Institute of Biochemistry, Royal Infirmary, Glasgow G4 0SF, UK

E. ROY SKINNER
Department of Molecular and Cell Biology, University of Aberdeen, Marischal College, Aberdeen AB9 1AS, UK

GERD UTERMANN
Institute of Medical Biology and Human Genetics, University of Innsbruck, Schöpfstrasse 41, A-6020 Innsbruck, Austria

PHILIP K. WEECH
Laboratory of Lipoprotein Metabolism, Institut de Recherches Cliniques de Montréal, 110 Pine Avenue West, Montréal, Québec, Canada H2W 1R7; present address: Merck Frosst Centre for Therapeutic Research, Merck-Frosst Canada Inc., CP/PO 1005, Pointe-Claire-Dorval, Québec, Canada H9R 4P8

PETER A. WILCE
Alcohol Research Unit, Department of Biochemistry, University of Queensland, Queensland 4072, Australia

Abbreviations

ABC	avidin–biotin horseradish peroxidase complex
ACAT	acylcoenzyme A:cholesterol acyltransferase
ACD	acid citrate dextrose
apo	apolipoprotein
APS	ammonium persulphate
ATCC	American Type Culture Collection
BHT	butylated hydroxy toluene
BSA	bovine serum albumin
cDNA	complementary deoxyribonucleic acid
CE	cholesteryl ester
CETP	cholesteryl ester transfer protein
CHO	Chinese hamster ovary (cells)
CoA	coenzyme A
CSF	cerebrospinal fluid
DAB	diaminobenzidine
DGU	density gradient ultracentrifugation
DiI	1,1'-dioctadecyl-3,3,3',3'-tetramethylindocarbocyanine
DiO	3,3'-dioctadecyloxacarbocyanine perchlorate
DMEM	Dulbecco's modified Eagle's medium
DMPC	dimyristoylphosphatidylcholine
DMSO	dimethylsulphoxide
DTNB	5,5'-dithiobis-(2-nitrobenzoic acid)
EAS	European Atherosclerosis Society
EDTA	ethylenediaminetetraacetate
EGTA	ethyleneglycol bis-(α-aminoethylether)-N,N,N',N'-tetraacetate
ELISA	enzyme-linked immunosorbent assay
FACS	fluorescence-activated cell sorter
FBS	fetal bovine serum
FCR	fractional catabolic rate
FCS	fetal calf serum
FH	familial hypercholesterolaemia
FPLC	fast protein liquid chromatography (Pharmacia)
G(L)C	gas (–liquid) chromatography
HAT	medium supplemented with hypoxanthine, aminopterine, and thymidine
HDL	high-density lipoproteins
Hepes	N-(2-hydroxyethyl)piperazine-N'-(2-ethanesulphonic acid)
HL	hepatic lipase
HMG	3-hydroxy-3-methylglutaryl
HPLC	high performance liquid chromatography
HPTLC	high performance thin layer chromatography
HSA	human serum albumin
IEF	isoelectric focusing

IDL	intermediate density lipoproteins
IgG	immunoglobulin G
IRMA	immunoradiometric assay
LCAT	lecithin:cholesterol acyltransferase
LDL	low-density lipoproteins
LPL	lipoprotein lipase
LPDS	lipoprotein-depleted (-deficient) serum
mAbs	monoclonal antibodies
mRNA	messenger ribonucleic acid
NCEPEP	National Cholesterol Education Program Expert Panel (on the Detection, Evaluation and Treatment of High Blood Pressure in Adults)
NIH	National Institutes of Health
PAGE	polyacrylamide (gradient) gel electrophoresis
PBS	phosphate-buffered saline
PC	phosphatidyl choline
PEG	polyethylene glycol
PMSF	phenylmethylsulphonyl fluoride
PTFE	polytetrafluoroethylene
PVDF	polyvinylidene difluoride
PVP	polyvinylpyrrolidone
R_F	rate of migration relative to the solvent front
$S_{0.5}$	50% of the maximal velocity
SAAM	simulation analysis and modelling
SDS	sodium dodecyl sulphate
S_F	flotation coefficient
TCA	trichloroacetic acid
TEMED	N,N,N',N'-tetramethylethylenediamine
TG	triglyceride (triacylglycerol)
TLC	thin-layer chromatography
TMU	tetramethylurea
Tris	tris(hydroxymethyl)aminomethane
UC	unesterified cholesterol
U/P	urine/plasma ratio
VLDL	very low density lipoproteins
WHO	World Health Organization

1

Lipoprotein separation and analysis for clinical studies

MICHAEL I. MACKNESS and PAUL N. DURRINGTON

1. Background

1.1 Introduction

It is fitting to open this volume by considering clinical investigation since much of the impetus for research into lipoproteins in recent years stems from their likely relevance to atherosclerotic vascular disease. A prime purpose of the lipoprotein system is to transport lipids, but its theme has undergone a variety of evolutionary variations in different animal species. It is becoming apparent that lipoproteins have wider biological significance than simply lipid transport and are involved in such diverse processes as immune reactions (1), coagulation (2), and tissue repair (3). At the same time it is realized that lipoprotein metabolism may follow very different pathways in different species and that observations made in animals are frequently inapplicable to man. There is thus very often no substitute for clinical studies in man. Lipoproteins were first observed by the medieval physicians when venesection was popular and the milky serum associated with diabetes mellitus, nephrotic syndrome, and overindulgence in alcohol was well described. However, the discovery that this appearance was due to circulating fat particles secreted by the gut into lymphatics after fat absorption was not made until 1771 by Hewson (4). There followed the finding that fat was also present in fasting serum (5), the discovery of cholesterol (6), and early descriptions of the various clinical syndromes associated with hyperlipidaemia (7).

Early studies of plasma lipids were interpreted as indicating a rather loose attachment of lipid to serum proteins, but it was not until the work of Macheboeuf (8) that it was realised that a discreet group of macromolecular complexes specifically for the carriage of lipids existed. Macheboeuf used a series of precipitation procedures, but the major work of identification of the lipoprotein classes was made by Gofman and coworkers (9) following the introduction of the analytical ultracentrifuge (*Figure 1*). The specific protein components of lipoproteins, the apolipoproteins, have come under intense biochemical scrutiny since the early 1960s.

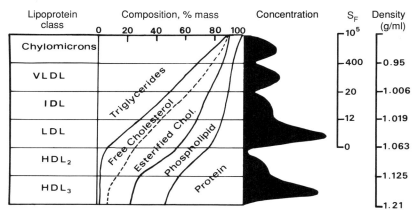

Figure 1. The classification, composition, and ultracentrifugation properties of the lipoproteins. VLDL, very low density lipoprotein; IDL, intermediate density lipoprotein; LDL, low density lipoprotein; HDL, high density lipoprotein; S_F, Svedberg flotation units.

The lipoproteins are macromolecular complexes of lipid and protein, a major function of which is to transport lipids through the vascular and extravascular body fluids. Great diversity of composition and physical properties are possible, particularly in disease, but also in health. As such, their classification and definition is particularly difficult. Each lipoprotein has a wide range of components, each with its own metabolic origin and fate. Lipoprotein components undergo a complex metabolic interplay with receptors, with enzymes located on the lipoproteins and on the capillary endothelium, and with other circulating lipoproteins, both in the vascular compartment and within the tissue fluid space (10). It is thus naive in the extreme to try to think of serum cholesterol or triglycerides in the same way as serum sodium or glucose, which are transported simply as solutes. The very existence of lipids within the circulation is dependent on lipoproteins.

1.2 Lipoprotein structure

The general structure of lipoprotein molecules is globular. The physico-chemical considerations, which govern the arrangement of their constituents, are similar to those involved in the formation of mixed micelles in the lumen of the intestine. Thus, within the outer part of the lipoprotein are found the more polar lipids, namely the phospholipids and free cholesterol, with their charged groups pointing out towards the water molecules. In physical terms, however, the role of the bile salts, which are also in the outer layer in the mixed micelle, is assumed by proteins, so that the surface of a lipoprotein structurally resembles the outer half of a cell membrane. Within the core of the lipoprotein particle are the more hydrophobic lipids, the esterified cholesterol, and the triglycerides. These form a central droplet to which are

2

anchored, by their hydrophobic regions, the surface coating molecules of phospholipid, cholesterol, and protein. The exception to this general structure is the newly formed or nascent HDL, which lacks the central lipid droplet and appears to exist as a disc-like bilayer, consisting largely of phospholipid and protein.

The protein components of lipoproteins are the apolipoproteins (*Figure 2*), a group of proteins of immense structural diversity, some of which have a largely structural role, others of which are major metabolic regulators and yet others of which may influence immunological and haemostatic responses apparently unconnected with lipid transport (*Table 1*). In addition, enzymes are found as components of lipoproteins. The leading example is lecithin: cholesterol acyltransferase (LCAT) which is located on the high density lipoproteins (HDL) which are also its site of action. Other enzymes with a less defined physiological role are also located on HDL, such as paraoxonase, which detoxifies organophosphorus compounds such as nerve gases and pesticides (11).

1.3 Lipid transport from liver and gut to peripheral tissues

The products of fat digestion (fatty acids, monoglycerides, lysolecithin, and free cholesterol) enter the enterocytes from the mixed micelles. They are re-esterified in the smooth endoplasmic reticulum of these cells. Long-chain

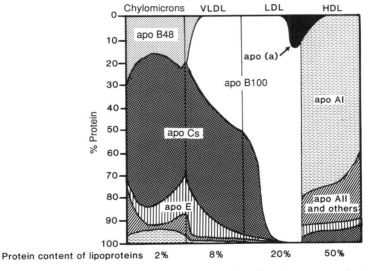

Figure 2. The distribution of the major apolipoproteins (apos) amongst the different lipoproteins. Each apolipoprotein is depicted as the proportion of the protein content of each lipoprotein which it comprises. The quantity of protein in each lipoprotein varies enormously. Thus, for example, more than 90% of the total apolipoprotein B100 in serum is present in LDL and more than 90% of total serum apolipoprotein AI is contained in HDL.

Table 1. Apolipoproteins: their properties and functions

Apolipoprotein	Molecular mass	Chromosomal location of gene	Plasma concentration (mg/100 ml) 3rd−97th percentile	Function
AI	28 016	11	80−160	Activation of LCAT Detergent properties
AII	17 414	1	20−55	? Activation of hepatic lipase
B48	264 000	2	0−2	Secretion of chylomicrons
B100	550 000	2	60−160	Secretion of VLDL, structural protein of LDL, receptor-mediated LDL catabolism
CI	6 600	19	3−11	?
CII	8 850	19	1−7	Activation of lipoprotein lipase
CIII	8 800	11	3−23	? Inhibition of hepatic uptake of chylomicrons and VLDL
E	34 100	19	2−6	Hepatic clearance of chylomicron remnants and IDL
(a)	300 000−700 000	6	1−100	?? Inhibitor of fibrinolysis

fatty acids (> 14C) are esterified with monoglycerides to form triglycerides and with lysolecithin to form lecithin. Free cholesterol is esterified by the enzyme, acyl-CoA:cholesterol O-acyltransferase (ACAT). The esterified lipids are then formed into lipoproteins.

The triglycerides, phospholipids, and cholesteryl esters are rapidly combined with an apolipoprotein, known as apo B48, produced in the rough endoplasmic reticulum of the enterocyte. The lipoproteins thus formed are further processed in the Golgi complex, where the apo B48 is glycosylated and actively transported to the cell surface for secretion into the lymph (chyle) and are termed chylomicrons. They are large (> 75 nm; density < 0.95 g/ml) and are rich in triglycerides, but contain only relatively small amounts of protein. They travel through the lacteals to join lymph from other parts of the body and enter the blood circulation via the thoracic duct. In addition to cholesterol absorbed from the diet, they may also receive cholesterol that has been newly synthesized in the gut or transferred from other lipoproteins present in the lymph and plasma. The newly secreted or nascent chylomicrons receive C apolipoproteins from the high density

4

lipoproteins (HDL), which in that respect appear to act as a circulating reservoir, since later in the course of the metabolism of the chylomicron, the C apolipoproteins are transferred back to the HDL pool. The chylomicrons also receive apolipoprotein E (apo E), although the manner in which they do so is unclear. Unlike other apolipoproteins, which are synthesized either in the liver or gut or both, apo E is exceptional in that it is synthesized (and perhaps secreted) by a large number of tissues: liver, brain, spleen, kidney, lungs, and adrenal gland. In part, apo E may come from HDL, but also it may be acquired directly as the chylomicrons circulate through the tissues.

Once the chylomicron has acquired the apolipoprotein, apo CII, it is capable of activating the enzyme, lipoprotein lipase. This enzyme is located on the vascular endothelium of tissues with a high requirement for triglycerides, such as skeletal muscle and cardiac muscle (for energy), adipose tissue (for storage), and lactating mammary gland (for milk). Lipoprotein lipase releases triglycerides from the core of the chylomicron by hydrolysing them to fatty acids and glycerol, which are taken up by the tissues locally. In this way the circulating chylomicron becomes progressively smaller. Its triglyceride content decreases and it becomes relatively richer in cholesterol and protein. As the core shrinks, its surface materials (phospholipids, free cholesterol, C apolipoproteins) become too crowded and there is a net transfer of these to HDL. The cholesteryl ester-enriched, triglyceride-depleted product of chylomicron metabolism is known as the chylomicron remnant. The apo B48, present from the time of assembly, remains tightly anchored to the core throughout. The apo E also remains and regions of its structure are exposed, which permit its catabolism via the 'remnant receptor' of the liver and also the apo B100/E receptors (often called LDL receptors), which can be expressed by virtually every cell in the body. It is possibly the case that apo E is inhibited from binding to its receptors earlier in the metabolism of chylomicrons, because its receptor-binding domain is blocked by the apolipoprotein, apo CIII. Remnants are largely removed from the circulation by the liver. Although the clearance of these particles via the apo B100/E receptor is theoretically possible, this route is not likely to contribute greatly to remnant uptake in the adult since the binding affinity of the hepatic apo E receptor for apo E is greater and the remnant particles must compete for binding at the apo B100/E receptor with low density lipoprotein (LDL), the particle concentration of which is much higher than that of the chylomicron remnants (even more so in the tissue fluid than in the plasma). Also the apo B100/E receptor is rapidly down-regulated by the lysosomal release of free cholesterol into the cell, which follows the entry of lipoprotein-receptor complexes into the cell, whereas expression of the remnant clearance pathway is unaffected by entry of cholesterol into the liver.

The liver itself secretes a triglyceride-rich lipoprotein known as very low density lipoprotein (VLDL). Teleologically, this allows the supply of triglycerides to tissues in the fasting state as well as postprandially. VLDL particles are somewhat smaller than the chylomicrons (30–45 nm in diameter; density < 1.006 g/ml). Once secreted they undergo exactly the same sequence of changes as chylomicrons; that is the acquisition of apolipoproteins and the progressive removal of triglycerides from their core by the enzyme, lipoprotein lipase. There are, however, some additional metabolic transformations involved in their metabolism in the human. In man, the liver, unlike the gut, does not esterify cholesterol before its secretion. This is different from species such as the rat. In the human, most of the cholesterol released from the liver each day into the circulation it secreted in the VLDL as free cholesterol and it undergoes esterification in the circulation. Free cholesterol is transferred to HDL along a concentration gradient. There it is esterified by the action of the enzyme, LCAT, which esterifies the hydroxyl group in the 3 position of cholesterol to a fatty acyl group. This it selectively removes from the 2 position of lecithin to give lysolecithin. The fatty acyl group in this position is generally unsaturated and the cholesteryl esters thus formed are frequently cholesteryl oleate or cholesteryl linoleate. Once formed, the cholesteryl ester is transferred back to VLDL. This cannot take place by simple diffusion because cholesteryl ester is intensely hydrophobic and because the concentration gradient is unfavourable. A special protein called cholesteryl ester transfer protein (CETP) or lipid transfer protein is present, which transports cholesteryl ester from HDL to VLDL. It does this in exchange for triglycerides in VLDL and thus also contributes to the removal of core triglycerides from VLDL. The major mechanism for the removal of triglycerides from VLDL is, however, the lipolysis catalysed by lipoprotein lipase.

Another major difference between VLDL and chylomicrons is that the apolipoprotein B produced by the liver in man is not apo B48, but is almost entirely apo B100. As in the case of the chylomicron, the quantum of apo B packaged in the VLDL remains tightly associated with the particle until its final catabolism and its amount does not vary after secretion. It is probable that each molecule of VLDL contains one molecule of apo B.

The circulating VLDL particles become progressively smaller as their core is removed by lipolysis and surface materials are transferred to HDL. In normal man most of the VLDL is converted to smaller low density lipoprotein (LDL) particles through the intermediary of a lipoprotein known as intermediate density lipoprotein (IDL). IDL has a density of 1.006–1.019 g/ml and possesses apo E. In this latter respect it is similar to chylomicron remnants. In some species, such as the rat, it is largely removed by the hepatic remnant receptor and LDL formation is thus bypassed. The enzyme, hepatic lipase, may be important in the conversion of IDL to LDL. In man LDL particles, which are relatively enriched in cholesterol, but are small enough (19–25 nm) to cross the vascular endothelium and enter the tissue fluid, serve to deliver cholesterol to the tissues. Their concentration in the extracellular fluid is probably about 10% of that in the plasma. The requirement for cholesterol is for cell membrane repair and growth and, in the case of specialized tissues such as the adrenal gland, gonads, and skin, as a precursor for steroid hormone and vitamin D synthesis. LDL is able to enter cells by two routes: one which is regulated according to the cholesterol requirement of each individual cell and one which appears to depend almost entirely on the extracellular concentrations of LDL.

The first of these two routes is by a cell surface receptor which specifically binds lipoproteins that contain apolipoprotein B100 or E. This is the apo B100/E receptor (LDL receptor). As has been mentioned previously, the receptor, although capable of binding apo E-containing lipoproteins, in practice usually binds largely to apo B100-containing lipoproteins of which LDL is the most widely distributed. After binding, the LDL-receptor complex is internalized within the cell where it undergoes lysosomal degradation. Its content of apo B is hydrolysed to its constituent amino acids and its cholesteryl ester is hydrolysed to free cholesterol. The release of this free cholesterol is the signal by which the cellular cholesterol content is precisely regulated by three coordinated reactions. The enzyme which is rate-limiting for cholesterol biosynthesis (3-hydroxy,3-methylglutaryl CoA reductase) is repressed, thus effectively centralizing cholesterol biosynthesis to organs such as the liver and gut. Secondly, the synthesis of the apo B100/E receptor itself is suppressed. Thirdly, ACAT is activated so that any cholesterol that is surplus to immediate requirements can be converted to cholesteryl ester, which because of its hydrophobic nature forms into droplets within the cytoplasm and is thus conveniently stored. The effect of lysosomal release of free cholesterol on the expression of the apo B100/E receptor contrasts with its effect on the hepatic remnant receptor which is not subject to any similar regulatory process. It is widely assumed, however, that although free cholesterol released by lysosomal digestion of cholesterol-rich apo E-containing lipoproteins entering the hepatocyte via the remnant receptor does not influence its own expression, it will, nevertheless, down-regulate the hepatic apo B100/E receptors.

Defective LDL uptake by the apo B100/E receptor is the basis of familial hypercholesterolaemia (see Section 5.4).

One other quantitatively important mechanism by which LDL cholesterol may enter cells is by a non-receptor-mediated pathway: LDL binds to cell membranes at sites other than those where the apo B100/E receptors are located and some of it passes through the membrane by pinocytosis. HDL is able to compete with LDL for this type of cell membrane association. The absence of a receptor means that the 'binding' is of low affinity and thus, at low concentrations, LDL entry by this route may have little significance. However, unlike receptor-mediated entry, non-receptor-mediated LDL uptake is not saturable, but continues to increase with increasing extracellular LDL concentrations. When LDL levels are relatively high, entry of cholesterol into cells by this route may thus assume greater quantitative importance than that via the apo B100/E receptor, which will be both saturated and down-regulated. This appears to be the situation in the adult human, whose LDL cholesterol is high relative to most animal species and in whom only about one-third of LDL is catabolized by receptors and two-thirds by non-receptor-mediated pathways. In hypercholesterolaemia, even more is catabolized via the non-receptor pathway (four-fifths in patients heterozygous for familial hypercholesterolaemia).

LDL may also be removed from the circulation by a number of receptors other than the apo B100/E receptor. Whilst these may be responsible for the catabolism of only relatively minor amounts of LDL, two such receptors present on the macrophage have excited considerable interest because they may lie at the heart of atherogenesis. They are the βVLDL receptor, a modified LDL receptor which *in vitro* allows the uptake of the βVLDL from patients with type III hyperlipoproteinaemia (Section 5.5) and the acetyl LDL receptor which permits the uptake of modified LDL by macrophages in tissue culture. Uptake at both these receptors is so rapid *in vitro* that foam cells resembling those in arterial fatty streaks are formed. On the other hand uptake of unmodified LDL by the macrophage via the apo B100/E receptor is too slow for foam cell formation and is even slower than for other cell types such as the fibroblast. Modifications which permit LDL uptake at the acetyl LDL receptor include acetylation (hence its name), but also oxidation of the LDL by cells in tissue culture and by redox cations such as copper. Such a phenomenon is of potential relevance to atherogenesis (12).

1.4 Transport of cholesterol from tissues back to liver

In the human, cholesterol is transported out of the gut and liver in quantities which greatly exceed its conversion to steroid hormones and its loss through the skin in sebum. Therefore, except when the requirement for membrane synthesis is high, for example during growth or active tissue repair, the greater part of the cholesterol transported to the tissues (if it is not to accumulate there) must be returned to the liver for elimination in the bile, or for reassembly into lipoproteins. The return of cholesterol from the tissues to the liver is termed 'reverse cholesterol transport'. It is less well understood than the pathways by which cholesterol reaches the tissues, but it may well be critical to the development of atheroma. HDL has many features which make it very likely that it is intimately involved in the reverse transport process.

The precursors of plasma HDL are probably disc-shaped bilayers composed largely of protein and phospholipid secreted mainly by the gut and liver which are converted to the spherical, mature form of HDL by the action of LCAT within the plasma compartment. HDL particles are also derived from surplus surface material (phospholipids, unesterified cholesterol, and apoproteins) of triglyceride-rich lipoproteins during lipolysis as described above. Apolipoproteins AI and AII comprise the major protein component of this nascent HDL. Its other apolipoproteins and the bulk of its lipid are acquired as it circulates through the vascular and other extracellular fluids. In this respect the transformation of HDL from its lipid-poor precursor to a relatively lipid-rich molecule is the opposite of that which the other lipoproteins undergo following their secretion.

HDL is a small particle compared with the other lipoproteins (5–12 nm) and easily crosses the vascular endothelium so that its concentration in the tissue fluids is much closer to its intravascular concentration than is the case for LDL. Because the serum HDL cholesterol concentration is only about one-fifth that of the LDL cholesterol concentration, it is often wrongly assumed that its particle concentration is lower. In fact, the particle concentrations of HDL and LDL in human plasma are often similar, and in the tissue fluids there are several times as many HDL molecules as those of other lipoproteins present unless capillaries have a fenestrated endothelium. Thus the cells are in contact with higher concentrations of HDL molecules than of any other lipoprotein. In man, unlike the rat, HDL serves no apparent function in transporting cholesterol to cells.

Recently it has been suggested that cells in tissue culture express receptors for HDL, particularly HDL_3 which might permit the transfer of cholesterol out of the cell. Passage across the cell membrane may not simply depend on receptors, however, since cholesterol must first be unesterified (unless CETP can cross the cell membrane to bind with it or possibly apo E synthesized within certain cells transports it out). Factors regulating the intracellular cholesterol esterase activity may also therefore be important. Once outside the cell, cholesterol must be re-esterified in order that it can be transported in any quantity in the core of lipoproteins. Therefore, whether or not HDL is involved as the initial acceptor molecule, cholesterol must at some stage on its return journey to the liver reside on HDL, because it is the site of LCAT activity. To add a further complication, however, cholesterol cannot simply be taken up by nascent HDL, be esterified and packed into its core, and then be cleared by the liver with the whole lipoprotein particle. This is because LDL equivalent to 1500 mg of cholesterol is catabolized each day, whereas the rate of catabolism of the HDL apolipoproteins AI and AII would permit less than 200 mg of HDL cholesterol to be catabolized each day. Therefore:

(a) the liver must be capable of selectively removing cholesterol from HDL and then returning the particle to the circulation with most of its apolipoproteins intact, or

(b) the cholesterol in HDL must be transferred to another lipoprotein class which is capable of being cleared in quantity by the liver, or

(c) a class of HDL which contains little apo AI or AII must be cleared by the liver at a much greater rate than the bulk of HDL.

In support of (a) there is some evidence that hepatic lipase might act on the phospholipid envelope of HDL during its passage through hepatic sinusoids, and release the cholesteryl ester contained in its core and that some hepatic trapping or even receptor-mediated mechanism might enhance the process. On the other hand, in support of (b) there is a well-established mechanism for the transfer of cholesteryl ester from HDL to VLDL through the agency of CETP. Once on VLDL, the cholesteryl ester can then arrive at the liver via the binding of IDL to remnant receptors or when LDL enters the liver after binding to the apo B100/E receptor or by the non-receptor-mediated route. Evidence for the third possibility, pathway (c), the return of cholesterol to the liver from HDL by a rapidly metabolized form of HDL present at low concentration in serum, is at present largely lacking.

It is incorrect to regard HDL as a single homogeneous species, since it is known to be a mixture of particles which differ in size, in lipid and apolipo-protein composition, and in function. Two peaks are seen in the analytical ultracentrifuge, the less dense of which is designated HDL_2 ($d = 1.063-1.125$ g/ml) and the more dense, HDL_3 ($d = 1.125-1.21$ g/ml). HDL_3 may be converted to HDL_2 by the acquisition of cholesterol, HDL_3 thus being a precursor form of HDL_2. Whereas antisera to apo AI precipitate virtually all of HDL, antisera to AII do not, suggesting that some molecules of HDL contain AI and AII, whereas others contain AI only. The AI-only HDL molecules which predominate in HDL_2 may arise from very different metabolic channels than the AI/AII particles. Furthermore, HDL may contain other molecular species with overlapping density ranges such as Lp(a) (see Section 5.2). HDL thus represents a rather heterogenous entity. (See Chapter 4 for a further discussion of HDL subfractions.)

2. Basic considerations of blood sampling

For the routine analysis of lipids and lipoproteins, blood should be taken in the fasting state (13) and therefore the quantitative and qualitative changes to lipoproteins which occur post-prandially are avoided. In our laboratory blood is withdrawn between 9.00 a.m. and 10.00 a.m. with patients lying semirecumbent after a fast from 10.00 p.m. on the day before. We also recommend that note be taken of the subject's general alcohol consumption, as this may acutely increase plasma concentrations of triglyceride-rich lipoproteins and HDL. There is considerable biological variation in lipid and lipoprotein levels which should always be taken into account when inter-preting results. In part this is due to seasonal changes, the menstrual cycle, and intercurrent illness, but there are other factors less well understood (13). Serum cholesterol levels will be within 10% of the true mean probably no more than 66% of the time. Fluctuations in triglyceride levels are even greater.

For the determination of total cholesterol and triglyceride concentration and of HDL cholesterol by precipitation, serum should be used and for the isolation of lipoproteins, plasma is preferred. The fasting venous blood should be collected into tubes containing no anti-coagulant for serum and tubes containing 30 mg EDTA disodium salt per 10 ml whole blood for plasma. The use of heparin as an anticoagulant should be avoided as it will interfere in lipoprotein separation by certain techniques. Allow the tubes to stand for 1 h at 4 °C. Then centrifuge them at 3000g for 30 min at 4 °C and pipette off the serum or plasma. Serum for lipid analysis may be frozen but serum for HDL quantitation by precipitation and plasma for lipoprotein isolation by other techniques should be stored at 4 °C before analysis. It is essential that plasma or serum for the analysis of lipoproteins is not frozen and additives should be used to prevent lipoprotein modification during storage (see Section 3.3.2). Isolation procedures should be undertaken within 3 days of storage at 4 °C (see Section 3.3.2).

3. The separation and quantitation of lipoprotein classes

3.1 Introduction

Several procedures exist to separate lipoproteins into their various classes. The techniques most commonly used in clinical research laboratories are precipitation, electrophoresis, and ultracentrifugation. These are considered in detail below.

3.2 Precipitation methods

Due to the specific interaction of apolipoprotein B (apo B) with a number of precipitating agents, lipoproteins containing apo B can be separated from non-apo B-containing lipoproteins by precipitation. Thus a mixture of VLDL and LDL can be separated from HDL (14). A number of precipitation methods are available for use (15–18). However, the two most popular are precipitation with heparin/manganese and precipitation with magnesium/phosphotungstic acid. Both these methods will be described here.

3.2.1 Sample preparation

Serum should be used for precipitation methods. If plasma is used twice the concentration of manganese is required for the heparin/manganese method (19) which creates even greater difficulties later if an enzymic method for cholesterol quantitation is used. Samples in which triglycerides are greater than 400 mg/100 ml (4.5 mmol/litre) must be pretreated to remove VLDL and chylomicrons before precipitation, particularly in the case of the heparin/manganese method. Failure to do so will result in their incomplete precipitation and hence to inaccurate HDL-cholesterol measurements. Chylomicrons and VLDL can be removed by ultracentrifugation for 18 h at 100 000g (20) or by ultrafiltration (21), but dilution is not recommended (17).

Protocol 1. Quantitation of high density lipoprotein cholesterol by magnesium/phosphotungstic acid precipitation

Materials

(a) 0.5 M MgCl$_2$
(b) 4% Phosphotungstic acid in 0.19 M NaOH

Procedure

1. To 0.5 ml of fresh serum (in duplicate) add 50 μl of MgCl$_2$ solution (a) and 50 μl of phosphotungstate (b).
2. Mix well and centrifuge immediately for 20 min at 2000g.
3. Remove the supernatant for measurement of HDL-cholesterol. At this stage the supernatant can be frozen prior to cholesterol measurement.

Protocol 2. Quantitation of high density lipoprotein cholesterol by heparin/manganese precipitation

Materials

(a) 2000 units/ml heparin (freeze-dried porcine intestinal mucous heparin)
(b) 0.55 M MnCl$_2$ (1.1 M, if EDTA plasma rather than serum used)

Procedure

1. Pre-cool 0.5 ml of freshly prepared serum to 4 °C and add 50 μl of heparin solution (a) and 50 μl of manganese chloride solution (b).
2. Mix well and incubate at 4 °C for 45 min.
3. Centrifuge for 30 min at 1600g, 4 °C. Remove the supernatant for HDL-cholesterol measurement. Supernatants can be stored frozen prior to cholesterol measurement.

If an enzymatic method is used on supernatants from the heparin/ manganese method, it is essential to incorporate EDTA in the reaction mixture and prevent the formation of a brown precipitate (22), which artefactually raises the apparent cholesterol concentration (see Section 4).

3.2.2 Choice of method

The heparin/manganese method was the method used in the Lipid Research Clinics Prevalence Program Study and is the method recommended by the International Federation of Clinical Chemistry. It gives results which are closest to those obtained by ultracentrifugation. The magnesium/ phosphotungstic acid method has the advantage of employing simple inorganic reagents which are easily obtained and stored, does not require incubation or a cooled centrifuge, is less affected by the presence of high concentrations of triglyceride-rich lipoproteins, and does not interfere with the enzymic determination of cholesterol (17). In the Lipid Research Clinics method which employs heparin/manganese, the triglyceride-rich lipoproteins are removed by ultracentrifugation and the infranatant HDL is then determined after precipitation of LDL, which overcomes the problem imposed by hypertriglyceridaemia but introduces the need for ultracentrifugation (see *Protocol 5*).

An unresolved problem with the magnesium/phosphotungstic acid method is that it gives lower values for HDL cholesterol, often by around 10%. This may be because it is particularly effective in precipitating the apo B-containing lipoprotein, Lp(a), present in the HDL-density range of many individuals (23), but it is also possible that it precipitates a variable amount of the apo AI-containing lipoproteins. Most commercial kits use the phosphotungstate/magnesium method and yet laboratories frequently base reference ranges on those obtained using the Lipid Research Clinics methodology (heparin/manganese). A concentration of HDL cholesterol of less than 35 mg/100 ml (0.9 mmol/litre) which is usually regarded as low (actually it is around the 10th percentile for the US population) would be less than 31 mg/100 ml (0.8 mmol/litre), if measured by the phosophotungstate/magnesium method. Epidemiological studies which have assessed HDL as a risk factor for coronary heart disease using the phosphotungstate/magnesium method have found it a weaker predictor than those employing heparin/manganese.

3.2.3 Friedewald formula (24)

When the serum triglycerides are less than 400 mg/100 ml (4.5 mmol/litre) the quantity of cholesterol in VLDL has a fairly constant linear relationship with total serum triglyceride concentration. This led to the development of a formula to permit the calculation of the LDL cholesterol from a knowledge of the HDL cholesterol and the concentrations of the total serum cholesterol and the triglycerides.

$$\text{LDL cholesterol} = \text{Total serum cholesterol} - \frac{(\text{HDL chol} + \text{total TG})}{5^*}\text{mg/100 ml}$$

* 2.2 if units are mmol/litre

This is enormously helpful in laboratories which do not possess an ultracentrifuge or when the size of a research study makes the cost of ultracentrifugation prohibitive. It has been suggested that a divisor of six rather than five for the triglycerides would be more accurate. Our own recent investigation does not support this suggestion (25) and our view of the other reported evaluations of the formula is that it has withstood the test of time surprisingly well. It should be emphasized that it cannot be applied to non-fasting serum or to patients with type III hyperlipoproteinaemia.

3.2.4 Quantitation of HDL subfractions

Although there has been a great deal of interest in the subfractionation of HDL into HDL_2 and HDL_3 in recent years, there has been remarkably little coordinated effort to develop laboratory methods which give consistent results between different laboratories and standardization has not even been contemplated. There is thus tremendous variation in the reported literature on the relative concentrations of HDL_2 and HDL_3 cholesterol in serum or plasma. Estimates vary from approximately equal amounts down to HDL_2 values of less than one quarter of those of HDL_3. These different estimates are largely method specific. Preparative ultracentrifugation tends to give relatively high HDL_2 cholesterol levels (26, 27). Estimates based on rate zonal ultracentrifugation and analytical ultracentrifugation are more prone to error and are thus more variable but generally these tend to confirm relatively high concentrations of HDL_2 (26, 28). On the other hand, dual precipitation methods using combinations of heparin/manganese and dextran sulphate (29) or polyethylene glycol (30) give lower levels of HDL_2 cholesterol and the lowest of all is with a method which involves ultracentrifugation of the heparin/manganese supernatant after conversion to a density of 1.125 g/ml (31). Our own experience with preparative ultracentrifugation and a dual precipitation method indicates that even greater divergence in results occurs in disease states such as diabetes and nephrotic syndrome as opposed to healthy people, presumably because the physical properties of HDL are altered and because of changes in Lp(a) concentration. Details of procedures for the separation of HDL subfractions are provided in Chapter 4.

3.3 Electrophoretic methods

3.3.1 Paper and agarose gel electrophoresis

Much of the early work in classifying the hyperlipoproteinaemias involved the use of paper electrophoresis and later agarose gel electrophoresis of plasma (32). These methods which are barely semiquantitative allow the identification of αHDL, βLDL, and preβVLDL bands, with chylomicrons remaining at the origin. They are still occasionally of value in the investigation of unusual hyperlipidaemias. A case has been proposed for their retention in the lipid laboratory as a screening test for type III hyperlipoproteinaemia (see Section 5.5). The broad β band in this condition, however, is frequently also produced in type IIb or V hyperlipoproteinaemia and so it is better to proceed to apo E phenotyping or ultracentrifugation when type III is suspected.

3.3.2 Polyacrylamide gradient gel electrophoresis (PAGE)

i. General considerations

Gradient gel electrophoresis separates lipoproteins according to their size. Lipoproteins migrate in an electric field through a gradient of increasing polyacrylamide concentration. The pore size of the matrix progressively decreases as the concentration of acrylamide increases. This produces a differential retardation of the migrating charged lipoproteins which are essentially separated according to their size. The migration of the particles effectively stops when they reach their exclusion limit. PAGE has found wide use in the identification of different apolipoproteins and apolipoprotein polymorphisms in delipidated lipoproteins. Descriptions of these techniques are to be found elsewhere in this book (Chapters 2 and 4).

ii. Methodological considerations (33)

Gels

The range of the acrylamide gradient will depend on the type of lipoprotein particles to be separated; generally a 4–30% gradient is effective for separating HDL while a 2–16% gradient is effective for separating LDL, IDL, and VLDL. It must be remembered that VLDL penetration is limited, generally to particles less than 40 nm in diameter. Gradient gels can be purchased premade from a number of suppliers (e.g. Flowgen, Bio-Rad).

Sample preparation

Gradient gel electrophoresis can be performed either on plasma directly or on lipoproteins isolated by ultracentrifugation. However, because separation is dependent on molecular size, steps must be taken during sample preparation to maintain the integrity of the particles. These include a number of additions to the plasma to prevent oxidative modification (EDTA 0.4% final concentration and storage under N_2), lipoprotein transformation (addition of 10 mM 5,5-dithiobis-(2-nitrobenzoic acid) to inhibit LCAT), protease breakdown (addition of 0.015% of phenylmethylsulphonylfluoride, PMSF), and bacterial decomposition (addition of 0.05% sodium azide). Prolonged storage of samples should be avoided. Under no circumstances should samples for gradient gel electrophoresis be frozen. High salt concentrations in lipoprotein fractions prepared by ultracentrifugation should be reduced by dialysis before electrophoresis.

Electrophoresis

After fixing the gel to the vertical slab gel apparatus, precondition the gel in electrophoresis buffer (90 mM Tris base, 80 mM boric acid, 3 mM sodium azide, 3 mM EDTA pH 8.3) for 20 min at 125 V, 10 °C. Dilute the sample (4:1 by volume) with a solution containing 40% sucrose and 0.05%

bromophenol blue. Apply the sample solution to the gel pocket as a volume equivalent to 10–15 μg of lipoprotein protein.

Electrophoresis is conducted by applying voltage to the gel in the sequence, 15 V for 15 min, 70 V for 20 min, and 125 V for 24 h. This is equivalent to 3000 V h.

Staining

The type of stain used is dependent on the sample. Protein staining should be used for isolated lipoproteins but when plasma has been electrophoresed, the large amounts of protein mask the lipoprotein bands therefore a lipid stain must be used:

(a) Protein staining—stain for protein by placing the gel in a solution of 0.05% Coomassie Blue R-250 in methanol:acetic acid:water (50:10:40, v/v) for 2 h and then destain in methanol:acetic acid:water (20:9:71, v/v) until the background is clear.

(b) Lipid staining—a number of general lipid stains are available but perhaps the most commonly used is Oil Red O. Stain the gels in a solution of 0.04% Oil Red O in 60% ethanol for a minimum of 24 h at 55–60 °C and destain in 5% acetic acid.

Calibration and particle size determination

Gels can be calibrated by applying a mixture of proteins with known hydrated diameters (see *Table 2*) alongside the samples. The mixture should contain 2.5 μg of each protein. Comparison of the migration distances of the calibration proteins with that of the lipoproteins will give an estimate of the particle diameter.

Densitometry

Densitometry of stained gels can be performed using standard commercially available equipment (Pharmacia, Bio-Rad). For gels stained with Coomassie Blue R-250, densitometry is performed at 555 nm; for Oil Red O stained gels a wavelength of 530 nm is used. Protein densitometry on lipid stained gels can

Table 2. Diameters of standard proteins for calibrating polyacrylamide gradient gel electrophoresis[a]

Protein	Hydrated diameter (nm)	Non-hydrated diameter (nm)
Thyroglobulin	17.0	11.6
Ferritin	12.2	10.2
Catalase	10.4	8.2
Lactate dehydrogenase	8.1	6.9
Bovine serum albumin	7.1	5.4

[a] Data from Nicohols *et al.* 1986 (33).

also be performed at 280 nm. By comparing the area under the densitometry peak of the standard proteins (2.5 μg) with those of the lipoproteins, an estimate of the amount of individual lipoproteins can be obtained. Computer-assisted densitometry can greatly assist both this estimation and that of the particle size.

Alternative methods for identification

Lipoproteins can be electrophoretically transferred to nitrocellulose sheets by the process known as western blotting. This procedure has been described in detail elsewhere (34) (see Chapter 3) and will not be described here. It may be noted that transfer of lipoproteins is less efficient than that of proteins, but can be increased by the addition of detergents to the blotting buffer which enhance the release of apolipoproteins from the lipoproteins. Specific apolipoproteins transferred to the nitrocellulose sheets are identified with specific antibodies and therefore the position of lipoproteins may be identified by their apolipoprotein content.

3.3.3 Isoelectric focusing

Lipoproteins can be separated according to the heterogeneity of their isoelectric points by the process of isoelectric focusing. Either polyacylamide or agarose gels can be used. At the moment, isoelectric focusing of whole lipoproteins has found little clinical application. The technique is, however, valuable in the separation of apolipoproteins and determining their poly-morphisms in delipidated lipoproteins and is described elsewhere in this book (Chapter 2). Of particular clinical importance is the identification of the apolipoprotein E isoforms (apo E2, E3, and E4). This permits apolipoprotein E phenotyping (35) which is relevant in the diagnosis of type III hyper-lipoproteinaemia (see Section 5.5), and the detection of apo CII which is helpful in the investigation of severe hypertriglyceridaemia (see Section 5 and *Table 7*).

3.4 Preparative ultracentrifugation

3.4.1 Introduction

Lipoproteins have lower hydrated densities than the other plasma proteins, permitting their isolation from plasma by flotation ultracentrifugation. This method can be used to prepare large amounts of lipoproteins for further investigation or small amounts for clinical studies. The density of plasma is increased by the addition of NaCl and/or KBr and during ultracentrifugation lipoproteins will float to the surface depending on their density and the prevailing small-solute density of the solution. The density range of plasma lipoproteins is given in *Figure 1*. Individual lipoprotein classes can be isolated by sequentially increasing the plasma density. Ultracentrifugation times tend to be long and it has been shown that during the ultracentrifugation, exchanges of lipid and apolipoprotein between lipoprotein classes occur: this is the principal disadvantage of this technique. However, since most of our current definitions of the lipoprotein classes rely on their hydrated density, ultracentrifugation remains the method of choice for many investigations. The three most important ultracentrifuge techniques employed in lipoprotein analysis are analytical ultracentrifugation (9, 36), zonal ultracentrifugation (37, 38), and preparative ultracentrifugation (using either fixed angle or swing-out rotors). Analytical ultracentrifugation and zonal ultracentrifugation are particularly specialized techniques and are outside the scope of this review.

i. Sample preparation
In order to prevent lipoprotein modification during ultracentrifugation the techniques described under electrophoresis should be performed. We recommend the addition of EDTA (0.04% final concentration), NaN_3 (0.05%), and PMSF (0.015%) to freshly prepared plasma to prevent lipoprotein modification.

3.4.2 Isolation of lipoproteins by sequential flotation ultracentrifugation in the fixed angle rotor

i. Introduction
These methods are very valuable for the preparation of lipoproteins for experimental purposes or for compositional analysis. Recovery may be low and operator skill and practice are required to give accurate levels of lipoprotein concentrations.

ii. Choice of rotors and centrifugation time
A number of rotors are available for lipoprotein separation and the choice depends largely on the type of ultracentrifugation and the number and volume of samples to be analysed. A fixed angled rotor minimizes the path the lipoproteins have to travel (39). The 60 Ti and 42.1 Ti Beckman rotors provide convenient examples to illustrate the technique. A convenient method for the measurement of lipoprotein concentrations for clinical purposes is described in *Protocols 3* and *4*. *Protocol 3* describes isolation of lipoproteins for laboratory studies, and *Protocols 4* and *5*, measurement of concentrations for clinical purposes.

Protocol 3. Method for the isolation of VLDL, IDL, LDL, and HDL

1. Treat 250 ml plasma as described under sample preparation (above).
2. Divide the plasma into polycarbonate thick walled centrifugation tubes (1 × 3.5 inches, 38.5 ml) and cap them. Balance with 0.15 M NaCl.
3. Centrifuge at 30 000 r.p.m. (100 000*g*), 10–15 °C for 20 h. Chylomicrons and VLDL float to the top of the tube leaving the other plasma proteins

and other lipoproteins at the bottom where a pellet containing lipoproteins and other proteins will form. If chylomicrons are to be removed separately, centrifuge the plasma at 20 000 r.p.m. for 30 min before the 20 h centrifugation.

4. Remove the layer of VLDL and some of the clear non-lipoprotein containing layer beneath this with a Pasteur pipette. (If highly purified VLDL is required, recentrifuge the pooled supernatant again under the same conditions. Also recentrifuge the infranatant if it is particularly important that this is uncontaminated with VLDL or chylomicrons. It should be remembered that during repeated ultracentrifugation some lipoprotein components are lost, particularly apo Cs.)

5. Resuspend the pellet in the infranatant and pool this solution. Measure its volume.

6. Raise the small solute (background) density of the infranatant solution by either of the following methods:

   (a) Add solid KBr. This has the advantage that the volume to be ultracentrifuged is only minimally increased. However, in practice this is rarely an advantage. The amount of KBr must be calculated making allowance for the fact that atomic volume is concentration and temperature dependent. Tables for this purpose are to be found in references 32 and 36.

   (b) The alternative method and the one favoured by the present authors is to add 'heavy solution' (40). Dissolve 153 g NaCl, 354 g of KBr, and 100 µg EDTA in 1 litre of water to give a density of approximately 1.33 g/ml, this should be regularly and carefully checked at 10–15 °C as some solute will tend to crystallize out on storage.

7. If IDL is the next lipoprotein fraction required, add sufficient heavy density solution to increase the density of the infranatant solution to 1.019 g/ml. The amount of heavy density solution to be added can be calculated by the following equation:

$$V_2 = V_1 \frac{D - D_1}{D_2 - D}$$

   where: V_1 = initial volume of solution, V_2 = volume of 'heavy density' solution to be added, D = required density, D_1 = original density, D_2 = density of the 'heavy solution'. (The small solute density of plasma and of 0.15 M saline is assumed to be 1.006 g/ml.)

8. After mixing, introduce the above solution of density 1.019 g/ml into the centrifuge tubes. This time it will be necessary to balance them by the addition of a stock solution of density 1.019 g/ml. A range of these should be made (e.g. 1.019, 1.063, 1.125, 1.21 g/ml) by the addition of

Protocol 3. *Continued*

calculated amounts of 'heavy solution' to 0.15 M saline (density 1.006 g/ml).

9. Centrifuge at 30 000 r.p.m. and 10–15 °C for 24 h. Remove the top layer of IDL in a small volume. Pool the remaining infranatants and increase the density to 1.063 g/ml with the appropriate volume of heavy density solution.

10. Introduce this solution into centrifuge tubes and balance using the 1.063 g/ml stock solution. Centrifuge at 30 000 r.p.m. and 10–15 °C for 28 h.

11. Remove the top layer of LDL in a small volume. Adjust the density of the pooled infranatant to 1.21 g/ml and centrifuge at 30 000 r.p.m. and 10–15 °C, for 48 h to float the HDL. Remove the HDL in a minimum volume. The above technique can be used to isolate single lipoprotein fractions in which case steps can be combined to reduce time. The total plasma lipoprotein fraction can, for example, be separated by increasing the density of plasma directly to 1.21 g/ml and centrifuging for 48 h.

12. If the isolated lipoproteins are to be further investigated by such techniques as gel electrophoresis, it is recommended that the high salt concentrations are removed by dialysis against a low salt buffer, e.g. 10 mm Tris–HCl, pH 8.0 containing 0.4% EDTA and 0.05% NaN_3 overnight at 4 °C. For some purposes, HDL should be dialysed before storage (but see also Chapter 4). It should be noted that the lipoproteins prepared in this way are contaminated with albumin. If necessary, this can be removed by recentrifuging the isolated lipoprotein fraction using the conditions described for the initial preparation.

3.4.3 Quantitation of plasma lipoproteins in the fixed angle rotor

The method described in the previous section would be suitable for the isolation of fairly large quantities of lipoproteins for experimental or analytical purpose. The recovery, however, is poor. When it is required to measure plasma lipoprotein concentrations for clinical purposes then a method giving greater recovery and smaller volumes is more suitable (see below). The 50.3 Ti Beckman rotor which holds 18 × 6.5 ml disposable polyallomer tubes (1/2 × 2 1/2 inch) is suitable.

Commonly it is required to use ultracentrifugation to remove lipoproteins of *d* less than 1.006 g/ml (chylomicrons and VLDL) so that a precipitation method may be used to measure HDL cholesterol in a sample which would otherwise be too lipaemic, as in the Lipid Research Clinics method (see Section 3.2.1 and *Protocol 2*). Frequently the cholesterol concentration of the lipoproteins of *d* less than 1.006 g/ml may be required for the diagnosis of type III hyperlipoproteinaemia (see Section 5.5). Perhaps, too, in a clinical trial or investigation it is required to measure VLDL cholesterol or triglyceride concentration.

Protocol 4. Quantitation of VLDL cholesterol by ultracentrifugation

1. Introduce 5 ml of plasma into a 6.5 ml ultracentrifuge tube and overlay with 1 ml of 0.15 M sodium chloride solution.
2. Balance the tubes carefully with a few drops more of this solution before ultracentrifugation, at 100 000g (40 000 r.p.m. in a 50.3 Ti rotor) and 10–15 °C for 18 h. This provides a good separation of the triglyceride-rich lipoproteins in the supernatant from the IDL, LDL, and HDL in the infranatant.
3. Remove the supernatant by pipetting or, better, by tube-slicing (Beckman Spinco Tube Slicer) and transferring it to a 5 ml volumetric flask (41). Return it to a volume of 5 ml by adding buffered 0.15 M saline before determination of its lipids.

Protocol 5. Quantitation of HDL cholesterol by ultracentrifugation

1. If HDL cholesterol is required to be measured in the ultracentrifuge, measure 3 ml of plasma into a 6.5 ml ultracentrifuge tube (0.5 × 2.5 inches) and add an amount of 'heavy solution' calculated from its measured density to give a density of 1.063 g/ml. Usually the volume of 'heavy solution' is approximately 2 ml.
2. Vortex the ultracentrifuge tube to mix the solutions and overlayer with about 1 ml of solution of density 1.063 g/ml.
3. After balancing with a few drops of 1.063 g/ml solution, ultracentrifuge the tubes at 40 000 r.p.m. at 20 °C for 48 h.
4. Remove the supernatant (VLDL, IDL, LDL) by pipetting or, better, by tube slicing (Spinco Tube Slicer, Beckman) and transfer the infranatant to a 5 ml volumetric flask. Return it to a volume of 5 ml by addition of 0.15 M buffered saline. Cholesterol determination will then give an HDL cholesterol concentration which is sufficiently accurate for most purposes. The HDL may be isolated by flotation at d 1.21 g/ml, but this entails further volume adjustment, and further ultracentrifugation and will considerably decrease the recovery and accuracy of the result.

For the quantitation of VLDL and HDL cholesterol we commonly employ a two tube technique which overcomes the difficulties and inaccuracy of sequential flotation. One tube is set up as described for d less than 1.006 g/ml, which in fasting subjects essentially yields VLDL, and the other as for the

HDL cholesterol, which yields an infranatant for HDL cholesterol and a supernatant containing VLDL and LDL combined. Cholesterol is also measured in the supernatant from this tube. LDL cholesterol can be calculated by difference and the recovery (usually > 95%) calculated from a knowledge of the total serum cholesterol (see Section 3.2.3).

The method may be made more sophisticated by, for example, including a measurement of HDL_3 cholesterol which entails a third tube being ultra-centrifuged at density 1.125 g/ml. To accomplish this, 2 ml of plasma are introduced into a 6.5 ml tube. The calculated volume of heavy solution (usually around 3 ml) to convert the small solute density to 1.125 g/ml is then added. After mixing, 1 ml of a solution of density 1.125 g/ml is overlayered and the tubes balanced with drops of the same solution. The infranatant obtained after ultracentrifugation for 48 h at 40 000 r.p.m. then contains the HDL_3 and the less dense lipoproteins will have floated clear as the supernatant. (See also Chapter 4.)

3.4.3 Density gradient ultracentrifugation

i. Introduction

The separation of lipoprotein classes by density gradient ultracentrifugation (DGU) in swinging bucket rotors has gained increasing popularity in recent years. The major advantage of this technique over the sequential ultra-centrifugation method for lipoprotein separation is a much reduced centri-fugation time. The volume of plasma that can be handled by the DGU technique is much less than with sequential ultracentrifugation and therefore, although this method is a powerful tool for the separation of lipoprotein classes, it is not the method of choice to produce large quantities of the lipoprotein classes. However, when clinical studies require a total profile of the lipoprotein classes, the increased speed of DGU may make it an attractive method.

DGU depends on floating the lipoprotein particles through a gradient formed by layering solutions of decreasing density (formed by NaCl, NaBr, and KBr) above a layer of plasma adjusted to a density greater than the solution above it. Centrifugation in a swinging bucket rotor causes the lipoprotein particles to float through the gradient. Two methods of DGU are available; one utilizes a continuous gradient and the other is a discontinuous gradient method. In continuous gradient DGU, the lipoprotein particles float to a particular part of the gradient dependent on their hydrated density. After ultracentrifugation, the gradient can be fractionated using a commercially available gradient fractionator or by punching a hole in the bottom of the centrifugation tube and collecting fractions. Both these techniques are time-consuming and require great care, as the gradient is easily disturbed (for example when the centrifuge tubes return from the horizontal to the vertical), resulting in a disturbance in the lipoprotein profile. In order to overcome these gradient fractionation difficulties, we routinely use discontinuous DGU and it is this method which will be described here.

ii. Discontinuous density gradient ultracentrifugation

A discontinuous gradient is prepared by layering solutions of successively lower densities above one another. This allows lipoprotein fractions of differing hydrated densities to be sequentially isolated as they float to the top of the gradient. The method described here (see *Protocol 6*) is based on that of Lindgren *el al.* (1972) (42) and is for use with the Beckman SW 40 Ti Swing-out rotor (which we use in conjunction with the Beckman LM-8 ultracentrifuge). Other manufacturers' rotors and ultracentrifuges can be used with the appropriate centrifugation times recalculated. Centrifugation tubes for the SW 40 Ti rotor can be obtained from a number of manufacturers. It is important when choosing tubes to ensure they are 'wettable', i.e. a smooth liquid flow can be obtained down the inner surface, so as to facilitate gradient formation. Non 'wettable' tubes can be coated with polyvinyl alcohol using the method of Holmquist (1982) (43) to ensure such a smooth liquid flow.

Solutions required

Two stock density solutions are required:

(a) $d = 1.006$ g/ml solution: 22.4 g NaCl, 0.2 g EDTA, and 2 ml of 1 M NaOH in 2 litres of distilled water.

(b) $d = 1.182$ g/ml solution: Add 249 g of NaBr to 1 litre of $d = 1.006$ g/ml solution.

The six solutions required to form the discontinuous gradient can then be prepared by mixing the two stock solutions in the proportions shown in *Table 3*.

Table 3. Preparation of solutions for discontinuous density gradient ultracentrifugation. Volumes of solution (a) and (b) are mixed to provide six solutions for the density gradient

Solution No.	Density (g/ml)	Solution (a) 1.006 g/ml (ml)	Solution (b) 1.182 g/ml (ml)
1	1.0988	75	83.67
2	1.0860	75	62.49
3	1.0790	75	53.16
4	1.0722	75	45.15
5	1.0641	75	36.93
6	1.0588	75	32.19

Sample preparation

To 2 ml of freshly prepared plasma add solid NaCl to give a density of 1.118 g/ml; 0.341 g NaCl is generally required, but for grossly hyperlipidaemic samples 0.35 g NaCl is used. Mix gently for 30 mins.

Protocol 6. Separation of lipoproteins by discontinuous density gradient ultracentrifugation

1. Carefully pipette 0.5 ml of $d = 1.182$ g/ml solution into the centrifuge tube and overlayer this with the 2 ml plasma sample.

2. Form the discontinuous gradient above the plasma by layering the six gradient solutions above the plasma (starting with solution 1). The amount of each solution to be added is given in *Table 4*. It is essential when pipetting the various solutions into the centrifuge tube to take great care to prevent mixing of the solutions. This can generally be achieved by pipetting the solution down the inside of the centrifugation tube (held at a 45 °C angle) with a slow continuous flow.

Table 4. Preparation of the discontinuous gradient

Solution	Density	Amount added[a] (ml)
1	1.0988	1
2	1.0860	1
3	1.0790	2
4	1.0722	2
5	1.0641	2
6	1.0588	2

[a] Total Volume including plasma = 12.5 ml

3. Centrifuge at 39 000 r.p.m. for 1 h 38 min at 23 °C to isolate lipoproteins S_F 60–400. Allow the rotor to come to rest *without* the aid of the brake and do not remove the tubes from the rotor buckets. Remove 1 ml of the S_F greater than 60 fraction from the top of the centrifuge tube using a fine bore Pasteur pipette (placed at the highest point of the meniscus) and transfer to a 1 ml volumetric flask. Once the sample has been removed, carefully layer 1 ml of $d = 1.0588$ g/ml (solution 6) on to the top of the gradient. Replace the rotor bucket and the rotor into the ultracentrifuge.

4. Centrifuge at 18 500 r.p.m. for 15 h 41 min. After centrifugation remove 0.5 ml (S_F 20–60) from the top of the gradient, replace the rotor bucket in the rotor and the rotor in the centrifuge.

5. Centrifuge at 39 000 r.p.m. for 2 h 35 min. Remove 0.5 ml (S_F 12–20) from the top of the gradient and replace rotor buckets and rotor.

6. Centrifuge at 30 000 r.p.m. for 21 h 10 min. Remove 1 ml (S_F 0–12) from the top of the gradient. After this the centrifugation is complete.

3.4.4 Single spin vertical density gradient ultracentrifugation

A recently developed alternative to classical DGU is single spin DGU in vertical rotors (44). This method also requires the creation of a salt gradient above the density adjusted plasma. Centrifugation takes place in a vertical rotor (a variety of which are available for Beckman and Sorvall centrifuges). The advantage of this method is its speed; separation of the lipoproteins can be achieved within 25–30 min depending on the type of centrifuge. Lipoproteins float to the appropriate portion of the gradient, according to their hydrated density. After centrifugation the gradient is removed from the tube with a gradient fractionator, a piece of apparatus which may not be readily available in routine clinical lipid laboratories. The major disadvantage of this methodology is wall adherence of VLDL and albumin which can account for losses of up to 40% of VLDL. Less plasma per rotor can be used in this method, than in sequential floatation which may determine the choice of method routinely adopted.

Single spin vertical DGU has been described in great detail elsewhere (44). A number of different protocols are described for optimum isolation of different lipoprotein classes and interested readers are recommended to refer to this source.

4. The measurement of serum and lipoprotein cholesterol and triglycerides

4.1 Introduction

Many methods are available for the measurement of cholesterol and triglycerides in serum and lipoprotein fractions. These include chemical methods (45) and methods based on the use of high-performance liquid chromatography and gas–liquid chromatography (46). These methods, however, tend to be lengthy and include the use of toxic chemicals or long extraction and clean up procedures. For the clinical laboratory, assays based on enzymatic techniques are probably more applicable (47, 48). They are simple to perform, rapid, easily automated, and available from a variety of manufacturers in simple kit form. Enzymic assays for both cholesterol and triglycerides are based on the generation of hydrogen peroxide from the substrate by a specific oxidase and the coupling of this through peroxidase to a chromogen. Production of the chromogen is stoichiometric and therefore the colour produced is directly proportional to the amount of analyte, which can be quantified by producing a calibration curve from known quantities of the analyte.

The following is typical of the reaction sequence used for cholesterol:

$$\text{Cholesteryl ester} + H_2O \xrightarrow{\text{Cholesterol esterase}} \text{Cholesterol} + \text{fatty acid}$$

$$\text{Cholesterol} + O_2 \xrightarrow[\text{oxidase}]{\text{Cholesterol}} \text{Cholesten-4-one} + H_2O_2$$

$$H_2O_2 + \text{4-Aminophenazone} + \text{Phenol} \xrightarrow{\text{Peroxidase}} \text{Quinoneimine} + H_2O$$

The quinoneimine dye (4-(*p*-benzoquinone-monoimino)phenazone) can be detected by its absorbance at 500 nm. Free cholesterol is measured by omitting the cholesterol esterase step. Kits for the determination of free and total cholesterol are available separately (from, e.g. Boehringer Mannheim, Biostat).

The following is typical of the reaction employed for triglycerides:

$$\text{Triglycerides} + 3\,H_2O \xrightarrow{\text{Lipase}} \text{Glycerol} + 3 \text{ fatty acids}$$

$$\text{Glycerol} + \text{ATP} \xrightarrow[\text{kinase}]{\text{Glycerol}} \text{Glycerol-3-phosphate} + \text{ADP}$$

$$\text{Glycerol-3-phosphate} + O_2 \xrightarrow[\text{oxidase}]{\substack{\text{Glycerol}\\\text{phosphate}}} \text{Dihydroxyacetonephosphate} + H_2O_2$$

$$2\,H_2O_2 + \text{4-Aminoantipyrine} + 4\text{ Chlorophenol} \xrightarrow{\text{Peroxidase}}$$
$$\text{Quinoneimine} + \text{HCl} + 4H_2O$$

It should be noted that glycerol is quantified. Thus conditions such as uncontrolled diabetes which cause hypertriglyceridaemia are a potential source of error. A blank which omits the lipase step would overcome this difficulty. Kits are available from Boehringer Mannheim, Biostat.

4.2 Practical points

(a) Each kit, whether for free or total cholesterol or triglyceride determinations, is accompanied by detailed manufacturers instructions. It is essential to follow these instructions exactly in order to ensure accuracy and reproducibility. Esterified cholesterol is usually calculated as the difference between the total and free cholesterol values.

(b) The range of linearity of the assays is limited and those samples which fall outside the calibration curve will require dilution with 0.15 M NaCl and the assay repeated. This is particularly true for hyperlipidaemic serum samples which are routinely diluted 1:3 for total cholesterol determinations and 1:4 for triglyceride determinations before assay.

(c) Cholesterol and triglyceride concentrations in separated lipoprotein fractions can be determined in exactly the same way as for serum, by

26

replacing serum with the fraction to be assayed in the method. Two minor problems arise when assaying for cholesterol and triglycerides in lipoprotein fractions:

i. Low levels of cholesterol and triglycerides in certain of these fractions, e.g. VLDL, may lead to inaccuracies in the determinations. To overcome this, the colour reaction can be enhanced by the addition of TBHBA. (2,4,6-tribromo-3-hydroxybenzoic acid). TBHBA is added to the cholesterol colour reagent to give a final concentration of 0.5% (w/v) and the reagent used as normal (49). A calibration curve using TBHBA should be run in addition to the samples.

ii. In the determination of HDL-cholesterol in HDL prepared by precipitation methods, the metal ions used in precipitation, particularly manganese, interfere in the cholesterol oxidase reaction (22). This can be overcome by the addition of EDTA to the cholesterol determination reagent. In our laboratory 2 ml of EDTA (4.27 g Na_2EDTA and 3.54 g Na_3EDTA in 100 ml distilled water) is added to 48 ml of cholesterol reagent immediately prior to using the reagent.

(d) Finally, in order to avoid day-to-day variation in the assays and to maintain accuracy and reproducibility it is essential that several quality control samples should be included in each batch of assays (1 per 10 or 15 assays). Quality control sera to be used as internal controls with known concentrations of cholesterol and triglycerides are available from a variety of manufacturers (e.g. Precinorm from Boehringer or Kontrollogen from Behring Diagnostics). Additionally, a number of external quality control schemes are available. Wellcome Diagnostics run a worldwide scheme. In the UK the National External Quality Assessment Scheme is run through the Wolfson Laboratories in Birmingham. In the USA a similar function is performed by the Centers for Disease Control. These periodically assess the performance of a laboratory through the supply of samples with unknown cholesterol and triglyceride concentrations.

4.3 Reproducibility and accuracy of lipid determinations

On both sides of the Atlantic the importance of lipid, and particularly cholesterol assays for the evaluation of clinical disorders has been emphasized (50–52). This is made more cogent by the introduction of guidelines for the clinical management of patients who have levels of serum cholesterol at the 75th percentile or less for the population so that small inaccuracies in the measurements make substantial difference to the placing of subjects in different therapeutic categories (see Section 5). It should always be remembered that biological variation may introduce much greater variability in the apparent serum cholesterol value than laboratory error and that important clinical decisions should not be based on single determinations (13).

5. Clinical evaluation of laboratory data

5.1 Consensus conference recommendations

5.1.1 Cholesterol

Serum cholesterol concentrations vary markedly between different individuals and the definition of normality poses considerable difficulties. The use of two standard deviations on either side of the mean to define a normal range is inappropriate since the median cholesterol level in populations with comparatively little ischaemic heart disease may be 2 mmol/litre or more less than that of societies like the UK where coronary disease is prevalent (10). Furthermore within populations such as the US or the UK, a 30-year-old man at the upper limit of such a range might for example have a ten times greater chance of a premature myocardial infarction than a similar person at its lower limit (53). In between these limits there is a progressive increase in risk with no sudden increase at any threshold cholesterol concentration even down to 4 mmol/litre (54). The relationship between coronary disease risk and cholesterol is, however, an exponential one, increasing progressively more steeply with increasing cholesterol levels. On the other hand the relationship between serum cholesterol concentration and all-cause mortality is not exponential but J-shaped (quadratic). The nadir for middle-aged men occurs in the range 175–200 mg/100 ml (4.5–5.2 mmol/litre), and this has led the European Atherosclerosis Society (EAS) to define 200 mg/100 ml (5.2 mmol/litre) as an optimal cholesterol level (55). Such a definition is provisional since evidence does not exist that lower cholesterol levels are the cause rather than the consequence of the increased mortality which is largely due to neoplasms, particularly of the large bowel (56). In the USA, therefore, the NIH National Cholesterol Education Program Expert Panel on the Detection Evaluation and Treatment of High Blood Cholesterol in Adults (NCEPEP) has already defined less than 200 mg/100 ml (5.2 mmol/litre) as desirable (57). In many societies, particularly those with a Northern European cuisine such as the USA, Canada, Australia, New Zealand, South Africa, and Northern Europe itself, the great majority of the population will have cholesterol levels exceeding 200 mg/100 ml (5.2 mmol/litre). Although, therefore, a valid objective of a public health approach to coronary prevention might be to reduce the average cholesterol to the desirable range, there would be many arguments against an entirely clinical approach. There is, for example, in 40-year-old men on average only about a two- to three-fold increase in coronary risk between cholesterol levels of 200 mg/100 ml (5.2 mmol/litre) and 270 mg/100 ml (7 mmol/litre). Such a small difference in risk would not justify therapeutic intervention with lipid-lowering drugs. However, if those individuals who do have heart attacks at an early age are regarded as having increased susceptibility by virtue of other risk factors such as cigarette smoking, high blood pressure, diabetes, low HDL cholesterol or genetic predisposition, then an attempt by the clinician to identify and treat such people is fully justified. Thus, for example, a 40-year-old man with a serum cholesterol value of 270 mg/100 ml (7 mmol/litre) who smokes and has moderate hypertension will have nine times the risk of a man with a cholesterol of 200 mg/100 ml (5.2 mmol/litre) without these other risk factors. There thus must be a higher level than the desirable or optimal level, where clinical awareness becomes important particularly when multiple coronary risk factors are present.

In defining such a clinical action limit, the level at which benefit from therapeutic intervention might ensue is important. Evidence for benefit from cholesterol reduction comes from two major primary intervention studies. In one, the Lipid Research Clinics Program Trial (58), the average cholesterol level before commencement of placebo or cholestyramine was 280 mg/100 ml (7.2 mmol/litre) and in the other, the Helsinki Heart Study (59) the cholesterol level before treatment with placebo or gemfibrozil was initiated was 270 mg/100 ml (7.0 mmol/litre). In angiographic studies of patients who have had previous coronary artery bypass grafts, evidence of decreased progression of disease in the native and grafted vessels was observed when even lower initial cholesterol levels were reduced to less than 200 mg/100 ml (5.2 mmol/litre) (60).

For these reasons the EAS has recommended that when serum cholesterol levels above 250 mg/100 ml (6.5 mmol/litre) persist despite dietary advice, lipid lowering drug therapy might be considered if other coronary risk factors are present (55). The authors (56) view the following as risk factors (these do not exactly coincide with the EAS or NCEPEP recommendations):

(a) genetic hyperlipidaemia
    i. familial hypercholesterolaemia
    ii type III hyperlipoproteinaemia

(b) adverse family history

(c) other coronary risk factors
    i. diabetes mellitus
    ii. hypertension
    iii. smoking history

(d) manifestations of coronary disease or peripheral arterial disease

(e) coronary artery bypass surgery

(f) hypercholesterolaemia combined with hypertriglyceridaemia and/or low HDL

(g) male sex

It is implicit in the EAS recommendations that when cholesterol is the only risk factor a higher level is required before it confers sufficient risk in its own right to justify drug therapy. It is also recommended that people whose cholesterol exceeds 300 mg/100 ml (7.8 mmol/litre) should be referred to a specialized lipid clinic. However the development of such clinics in countries such as the UK is at present inadequate to allow this.

5.1.2 LDL cholesterol

The NCEPEP recommends that LDL cholesterol rather than total serum cholesterol is used to define the category of risk (57). Thus serum LDL cholesterol should be measured in all people whose total cholesterol exceeds 240 mg/100 ml (6.2 mmol/litre) (Table 5) and also in those whose total cholesterol is greater than 200 mg/100 ml (5.2 mmol/litre) if they have pre-existing evidence of ischaemic heart disease or they have one additional risk factor in the case of a man or two in a woman. People whose LDL cholesterol exceeds 130 mg/100 ml (3.3 mmol/litre) are given dietary treatment. This level is equivalent on average to a total cholesterol of 200 mg/100 ml (5.2 mmol/litre). The recommendations are thus similar to the European ones. There are, however, much more detailed dietary recommendations and a two-step diet is recommended depending on the cholesterol response. NCEPEP advises consideration of lipid-lowering drugs if the LDL cholesterol persists at levels above 190 mg/100 ml (4.9 mmol/litre) (equivalent on average to a total cholesterol of 270 mg/100 ml or 6.9 mmol/litre) or more than 160 mg/100 ml (4.1 mmol/litre) (equivalent on average to a total cholesterol of 240 mg/100 ml or 6.5 mmol/litre) if coronary disease is already present or there is at least one other risk factor in a man or two or more other risk factors in a woman. Thus the NCEPEP define a level where drugs might be used in the absence of other risk factors. Although the NCEPEP recommend that decisions be based on the LDL cholesterol, in other respects the levels they define are broadly equivalent to the European ones.

Table 5. Recommendations of the European Atherosclerosis Society (EAS) (55) and the National Cholesterol Education Program Expert Panel (NCEPEP) (57), for the clinical evaluation of serum cholesterol and LDL cholesterol levels

	Cholesterol mg/100 ml (mmol/litre)	
	EAS recommendation	**NCEPEP recommendation**
Desirable	TC^a < 200 (5.2)	TC < 200 (5.2)
Diet	TC > 200 (5.2)	LDL^b > 130 (3.3)
Consider drugs if despite diet:	TC > 250 (6.5)	LDL > 160 (4.1) if other risk factors present, > 190 (4.9) if not

^{a} TC, total serum cholesterol.
^{b} LDL, low density lipoprotein cholesterol concentration.

Both sets of recommendations are based on information about cholesterol and risk which was largely obtained in studies which employed chemical methods to measure cholesterol. Enzyme assays may introduce a bias (50–52). Furthermore the recommendations based on LDL cholesterol assume the use of the Friedewald formula in conjunction with the heparin/manganese method for HDL cholesterol. As was previously discussed (Section 3.2.2), the phosphotungstate/magnesium method which gives lower estimates of HDL cholesterol is commonly used in clinical laboratories, which means that higher LDL values will result. Also many clinical laboratories have cholesterol assays which perform badly in the range required for HDL cholesterol measurement and standardization and quality control of HDL measurement is largely non-existent. Thus the measurement of LDL cholesterol may introduce imprecision rather than further sophistication into patient management once it strays outside the research laboratory.

It should be noted that, although good data exist for the changes in total serum cholesterol, triglyceride, and lipoprotein cholesterol with age and differences between men and women in the US population (61) and also reasonably comprehensive data exist for serum cholesterol in the UK (62–65), neither the EAS or NCEPEP recommendations take these factors into account. It should, however, be remembered when assessing risk in the individual patient that the predictive power of serum cholesterol and LDL cholesterol is much less in older age groups and that there is a dearth of clinical trial evidence for benefit from cholesterol reduction in women of any age and in both men and women over the age of 65 years.

5.1.3 Triglycerides

The 95th percentile for fasting serum triglycerides in middle-aged men is approximately 300 mg/100 ml (3.4 mmol/litre) and this was regarded by the NCEPEP as high. The EAS recommended that significance be attached to levels of around 200 mg/100 ml (2.3 mmol/litre). For neither of these levels is there evidence in non-diabetic subjects that they act as an independent risk factor in multivariate analysis. However, their association with low serum HDL cholesterol, hyperapobetalipoproteinaemia, increased intermediate density lipoprotein, hyperfibrinogenaemia, and glucose intolerance (10) explains their strength as a risk factor on univariate analysis and makes TG concentration a useful adjunct to other risk factors in the clinical evaluation of cardiovascular risk. The presence of raised serum triglycerides increases the risk of raised cholesterol still further. In the absence of elevated cholesterol concentration, the use of lipid lowering therapy for hypertriglyceridaemia *per se* is difficult to justify unless the levels are high enough to suggest that acute pancreatitis might occur. Acute pancreatitis does not generally ensue unless levels of triglycerides exceed 3500 mg/100 ml (40 mmol/litre) and even then many individuals are resistant (10). However, patients with fasting triglyceride values above 450 mg/100 ml (5 mmol/litre) may under some circumstances (e.g. a large fatty meal, high alcohol intake, administration of a β-adrenoreceptor blocking drug or oestrogen) develop levels high enough to provoke acute pancreatitis. Certainly fasting triglyceride levels above 1000 mg/100 ml (11 mmol/litre) require careful management. There is no reliable evidence that suggests that persistently high triglycerides at any level lead to chronic pancreatitis other than through the agency of recurrent acute pancreatitis.

5.1.4 HDL cholesterol

Serum HDL cholesterol levels below the 10th percentile should be regarded as an additional risk factor in the management of hypercholesterolaemia (56, 57). This probably equates to 35 mg/100 ml (0.9 mmol/litre) using the heparin/manganese method and 31 mg/100 ml (0.9 mmol/litre) by phosphotungstate/magnesium.

5.2 Lipoprotein (a)

This lipoprotein is present in the same density region as LDL and HDL_2 (see *Figure 2*), its precise location varying from individual to individual (2). Its concentration also shows great individual variation from virtually undetectable amounts to levels exceeding 1000 mg/100 ml. Like LDL, the other major component of the protein moiety of Lp(a) is apolipoprotein B. Lipoprotein (a) (Lp(a)) differs from LDL in uniquely having an apolipoprotein designated apo(a). This is probably directly disulphide-linked to apo B. Apo(a) is a huge mutant form of plasminogen in which its kringle 4 domain is repeated many times. The number of these repeats determines its molecular weight which may vary from 300 000–700 000 Da. Those individuals with the highest levels of Lp(a) exhibit the smaller molecular weight polymorphisms and this explains to a large extent the strong influence of inheritance on the serum Lp(a) concentration.

Lp(a) is usually quantitated by immunoassay of its apo(a) component (see Chapter 2), although the mass units which are frequently used to express its concentration relate to the total protein (apo B and apo(a)) and do not take into account the variation in its molecular weight. Immunoelectrophoresis and radial immunodiffusion are too insensitive to be regarded as suitable assay methods for Lp(a). Recommended methods are thus immunoradiometric assay (IRMA) and enzyme-linked immunosorbent assay (ELISA). Also methods which do not employ monoclonal antibodies are unsuitable because of cross reaction with plasminogen.

It is often said that Lp(a) is a strong predictor of coronary risk. The evidence suggests it is rather weak (2). However, in the presence of raised LDL levels (either measured as LDL cholesterol or apo B) it appears to increase the likelihood of coronary disease occurring at any early age rather more dramatically (66, 67). More prospective studies are required since much of our information, as with apolipoproteins AI and B, is based on case-control studies. However, levels of greater than 20 mg/100 ml would increase the risk from LDL cholesterol levels above 200 mg/100 ml (5.2 mmol/litre) or apo B levels greater than 130 mg/100 ml.

There are at present no standards for Lp(a). This is largely because of lack of agreement and lack of familiarity with apolipoproteins amongst clinical chemists. Lp(a) may not prove as great a challenge as apo B, since its apo(a) can be isolated in aqueous solution if the disulphide bonds in Lp(a) are reduced. If the starting material is plasma from homozygotes for the different polymorphisms then the problems of variable molecular weight can also be addressed.

5.3 WHO classification of hyperlipoproteinaemia

The hyperlipoproteinaemias were defined by Fredrickson, Levy, and Lees in a series of articles in 1967 (68). Five phenotypes were described according to which of the lipoproteins was increased, and later, type II was subdivided into IIa and IIb (69) (*Table 6*). The phenotypes are frequently and wrongly regarded as a diagnostic classification. It cannot be overemphasized that this is not the case. It is often no more than a way of reporting which serum lipoprotein is increased. All of the types may be either primary or secondary. Important secondary causes are obesity, drugs (e.g. beta blockers, thiazide diuretics, oestrogens), diabetes mellitus, alcohol, renal disease, hypothyroidism, biliary obstruction, myeloma, and pregnancy (71). Even within the primary types there are recognisably different disorders. Thus, for example, primary type IIa and occasionally type IIb may occur as the result of a single defective gene producing the clinical syndrome of familial hypercholesterolaemia with such manifestations as tendon xanthomata and a substantially decreased life expectancy due to coronary heart disease (see later). Even more commonly, however, the type IIa or IIb phenotype results from the interaction of at least two genes (polygenic) and environmental influences, particularly a high dietary intake of fat. This probably never leads to the formation of tendon xanthomata and in many individuals has a more favourable prognosis than familial hypercholesterolaemia. As more of the molecular mechanisms underlying the different phenotypes are uncovered, it is increasingly realised that they represent a heterogeneous group of different diseases.

The WHO phenotype can be established in the great majority of patients without redress to any more sophisticated techniques than visual inspection of serum and the determination of fasting levels of serum cholesterol and triglycerides. The determination of HDL cholesterol is a further refinement which allows the calculation of LDL cholesterol (see 3.2.3). To be consistent with the US and European guidelines, LDL cholesterol should be regarded as elevated when it exceeds 175 mg/100 ml (4.5 mmol/litre) (i.e. midway between 160 mg/100 ml and 4.1 mmol/litre with CHD or two other risk factors or 190 mg/

Table 6. The classification of the hyperlipoproteinaemias

WHO classification	Lipoprotein elevated (fasting)	Prevalence	Primary causes	
			Established	**Putative**
IIa	LDL	Common	Familial hypercholesterolaemia (LDL receptor defect)	Polygenic Sporadic Familial combined hyperlipidaemia
IIb	LDL + VLDL	Common	Familial hypercholesterolaemia (less likely than (in IIa)	Polygenic Sporadic Familial combined hyperlipidaemia (more likely than in IIa or IV) Apo B3500 mutation
IV	VLDL	Common	None	Polygenic Sporadic Familial combined hyperlipidaemia Hyperapobetalipo-proteinaemia Familial hyper-triglyceridaemia[a]
III	ß VLDL	Unusual	Apo E2 homozygosity[a] other apo E mutations producing defective remnant uptake Apo E2 deficiency	
V	Chylomicrons + VLDL	Unusual	Familial lipoprotein lipase deficiency Apo CII deficiency	Familial hyper-triglyceridaemia[a] Polygenic Sporadic[a] Familial combined hyperlipidaemia[a]
I	Chylomicrons	Rare	Familial lipoprotein lipase deficiency[b] (especially in childhood and not on low fat diet) Apo CII deficiency[b]	

[a] Usually combined with some additional factor provoking hypertriglyceridaemia.
[b] Reference 70.

100 ml and 4.9 mmol/litre without CHD or two other risk factors). This equates approximately to a total serum cholesterol exceeding 250 mg/100 ml (6.5 mmol/litre). We have tended to continue to accept a definition of hypertri-glyceridaemia for triglyceride levels greater than 180 mg/100 ml (2.0 mmol/litre), accepting that this does not in itself indicate any therapeutic response.

Thus in type IIa, LDL cholesterol exceeds 175 mg/100 ml (4.5 mmol/litre) and fasting serum triglycerides are less than 180 mg/100 ml (2.0 mmol/litre). In type IIb, both LDL cholesterol and triglycerides are raised. To identify the IIb phenotype, type V hyperlipoproteinaemia must be excluded by inspection of the serum for milkiness and, if in doubt, left to stand when a creamy layer will form on the surface if chylomicrons are present. Type IIb is unlikely to occur when triglyceride concentrations exceed the cholesterol concentration or when triglycerides are in excess of 900 mg/100 ml (10 mmol/litre). When triglycerides exceeds 500 mg/100 ml (4.5 mmol/litre), the Friedewald formula is inaccurate and total cholesterol does not necessarily reflect an increase in LDL cholesterol since the triglyceride-rich lipoproteins (VLDL and chylomicrons) will contain a relatively larger proportion of total serum cholesterol. When both cholesterol and triglycerides are raised and their molar concentrations are very similar (the mass of triglycerides is about twice that of cholesterol) then type III hyperlipoproteinaemia should be considered (see Section 5.5).

Type IV occurs when LDL cholesterol levels are normal and serum cholesterol is commonly below 250 mg/100 ml (6.5 mmol/litre). Fasting serum triglycerides are raised, but the serum is opalescent in indirect light rather than milky, and their concentration will generally not exceed 900 mg/100 ml (10 mmol/litre). In the rare circumstance where much higher levels of triglycerides occur with a relatively normal cholesterol the serum will be milky and type I hyperlipoproteinaemia is possible. On standing, chylomicrons will form a creamy layer on the surface, but the serum underneath will be clear rather than opalescent since there is no increase in VLDL.

5.4 Familial hypercholesterolaemia

Familial hypercholesterolaemia (FH) is the most common clinical syndrome leading to premature coronary heart disease which we are currently able to identify. It is dominantly inherited and heterozygotes occur in the US and UK population with a frequency of about 1 in 500 (10). Some 80% of men and 50% of women with heterozygous FH will have symptomatic coronary heart disease before the age of 60 years if untreated (10, 72). The mortality in the men is greater than 50% and in women 15% by the age of 60 years. The WHO phenotype is usually type IIa and occasionally IIb (which carries a rather worse prognosis). Serum cholesterol levels in adults with the condition usually exceed 350 mg/100 ml (9 mmol/litre). Being an entirely genetic syndrome, elevated levels of cholesterol are found throughout childhood and there are good reasons for commencing at least dietary therapy and other coronary prevention measures in childhood. Some children have clearly elevated cholesterol levels (275 mg/100 ml or 7 mmol/litre) and when a parent is known to have the syndrome the diagnosis is not difficult. However, this is not always the case, for example at adolescence serum cholesterol levels decline and also in earlier childhood when the cholesterol is more responsive to diet than in adults, a child from a coronary prone family may well be on a low saturated fat diet when blood is first taken, obscuring the hypercholesterolaemia. Thus levels of cholesterol greater than 230 mg/100 ml (6 mmol/litre) should be viewed with suspicion and will need to be repeated as the child matures.

The clinical hallmark of heterozygous FH in adults is the presence of tendon xanthomata. These are typically found in the Achilles tendons, where they may cause attacks of tenosynovitis. They also frequently occur in the tendons on the dorsum of the hands. In both locations the xanthomata are deep within the tendons and thus do not appear yellow through the skin. They are hard because of the presence of fibrous tissue and those on the dorsum of the hand may have the consistency of bone. Occasionally subperiosteal xanthomata occur at the tibial insertion of the patellar tendon and more rarely in other tendons. The definition of heterozygous familial hypercholesterolaemia adopted by the Simon Broome register (10) is:

Definite familial hypercholesterolaemia

(a) Cholesterol level above 260 mg/100 ml (6.7 mmol/litre) in children under 16 or 290 mg/100 ml (7.5 mmol/litre) in an adult or LDL level above 190 mg/100 ml (4.9 mmol/litre) in adults.

Plus

(b) Tendon xanthomata in patient or in first or second-degree relative.

Possible familial hypercholesterolaemia

(a) As above plus one of (c) or (d)

(c) Family history of myocardial infraction below age of 50 in second-degree relative or below the age of 60 in first-degree relative.

(d) Family history of raised cholesterol levels above 290 mg/100 ml (7.5 mmol/litre) in first- or second-degree relative.

The cause of familial hypercholesterolaemia is defective receptor-mediated LDL catabolism. This is generally due to a defective LDL receptor gene which results either in impaired receptor function or failure of its synthesis (72–74). An enormous number of different mutations can occur in the syndrome making a DNA marker for FH generally applicable outside individual kindreds an unlikely possibility. Hence our continuing reliance on clinical features for the diagnosis. A defect of apo B affecting its receptor binding site could produce a similar clinical syndrome. One mutation which influences receptor binding is the apo B3500 mutation. However, this is much less commonly the cause of the FH syndrome than defects in the LDL receptor mechanism itself (75).

FH homozygotes are uncommon as would be expected since only one in 250 000 marriages are likely to occur between heterozygotes and then only one in four of the children of such marriages would be homozygotes. The condition is severe in homozygotes; these subjects develop coronary heart disease in childhood and develop extensive subcutaneous planar xanthomata and polyarthritis at an even earlier age. With many possible mutations of the receptor gene, most homozygotes are really mixed heterozygotes unless they arise in a closed inbred population. There is considerable variation in the expression of homozygous FH. Few homozygotes, however, have the capacity to survive beyond the age of 20 years, the worst combination being the combination of two mutations that result in a failure of receptor expression rather than a defective receptor. The serum cholesterol generally exceeds 600 mg/100 ml (15 mmol/litre).

5.5 Type III hyperlipoproteinaemia

This is a clinical syndrome in which hyperlipidaemia is combined with the presence of an abnormal cholesterol-rich VLDL which has β mobility on electrophoresis (VLDL) (10, 74, 76). It is much less common than heterozygous FH, probably affecting fewer than one in 10 000 people. It is associated with a high risk of coronary artery disease (probably almost as great as that of heterozygous FH), but in addition there is a pronounced tendency to develop peripheral arterial disease, particularly of the femoropopliteal arteries. It produces striate palmar xanthomata and tuberose and tubero-eruptive xanthomata. Although classically associated with a broad β band on electrophoresis, this band is rather non-specific, since similar electrophoretic appearances are sometimes seen with type IIb and V hyperlipoproteinaemia.

At least 75% and probably more than 90% of patients with type III hyperlipoproteinaemia are homozygotes for apo E2. In patients in whom the diagnosis is suspected polyacrylamide isoelectric focusing of delipidated VLDL (see Chapter 2) is thus a useful diagnostic test (35, 77). It should be noted that around 1% of the population of Europe and the USA are apo E2 homozygotes and that in the absence of hyperlipidaemia this does not indicate any clinical diagnosis (indeed their serum cholesterol is generally lower than average). The mutation in apo E2 is the substitution of cysteine for the arginine residue which occurs as the 158th amino acid in apo E3 and E4. It is generally considered that some other stimulus to hyperlipidaemia, such as obesity, diabetes, hypothyroidism or familial hypertriglyceridaemia in addition to apo E2 homozygosity, is required to produce the type III phenotype. Thus it is usually an autosomal recessive condition with variable penetrance. However, occasionally other mutations of the apo E gene produce a much stronger tendency to type III hyperlipoproteinaemia. Thus the substitution of cysteine for arginine at position 142 or the insertion of seven additional amino acids in the region 121–127 (apo E_{Leiden}) may result in dominant expression. Type III hyperlipoproteinaemia is exceedingly rare in women before the menopause.

The laboratory diagnostic criterion for the diagnosis of type III hyperlipoproteinaemia proposed by the NIH group on the basis of a large series if trials is one which most centres follow. This requires the isolation of VLDL in the ultracentrifuge (see Section 3.4.3). βVLDL is considered to be present if the concentration of VLDL cholesterol expressed as a ratio of the total serum triglyceride concentration exceeds 0.3 (if mg/100 ml are used) or 0.68 (if mmol/litre are units used) (78). Ratios greater than 0.25 or 0.57 respectively are highly suspicious. The procedure should not be applied unless there is hyperlipidaemia present, because some normal people have similar ratios. On the other hand false low ratios can be found in the occasional type III patient who also has marked chylomicronaemia (seen most often in diabetics with type III). Under these circumstances, the test should be repeated after several days on a 20 g fat diet. It should also be remembered that sometimes when myeloma produces hyperlipidaemia the VLDL cholesterol concentration may be elevated into the type III range.

5.6 Hypolipoproteinaemias

Occasionally in the clinical laboratory serum samples are encountered with abnormally low serum cholesterol levels. This is increasingly the case as population screening is undertaken. Low total serum cholesterol is invariably due to low levels of LDL and when profound there may also be a decrease in HDL cholesterol. There are primary causes of LDL deficiency such as abetalipoproteinaemia and hypobetalipoproteinaemia (10, 79). However, before entertaining these diagnoses care should be taken to exclude other causes of LDL deficiency such as malignancy (for example rectal or prostatic neoplasms, leukaemia, myeloma etc.) and malabsorption (short bowel, blind-loop syndrome, coeliac disease, pancreatic exocrine insufficiency, giardiasis, etc.) (steatorrhoea is, of course, a feature of abetalipoproteinaemia, but other causes of malabsorption must still be considered). Low levels of LDL as well as HDL occur in Tangier disease (analphalipoproteinaemia), but this is exceedingly rare.

When low HDL levels are encountered it is commonly in association with hypertriglyceridaemia, either primary or secondary and/or obesity. Other secondary causes such as liver disease and renal disease should be excluded. Familial lecithin:cholesterol acyltransferase deficiency, Tangier disease, and Fish-eye disease should sometimes be considered (10, 78) but are unlikely. The causes of more commonly encountered low HDL levels are at present poorly differentiated and specific clinical diagnoses are often not possible.

Acknowledgements

MIM is supported by the North Western Regional Health Authority. Some of the work referred to here was supported by them and the Medical Research

Council. The authors would like to thank Miss C. Price for her patience in typing this manuscript.

References

1. Harmony, J. A. K., Aleson, A. L., McCarthy, B. M., Morris, R. E., Scupham, D. W., and Grupp, S. A. (1986). In *Biochemistry and Biology of 'Plasma Lipoproteins'* (ed. A. M. Scanu and A. A. Spector) p. 403. Marcel Dekker, New York.
2. MBewu, A. and Durrington, P. N. (1990). *Atherosclerosis*, **85**, 1.
3. Mahley, R. W. (1988). *Science*, **240**, 622.
4. Hewson, W. (1771). *An Experimental Enquiry into the Properties of the Blood with Remarks on Some of its Morbid Appearances and an Appendix Relating to the Discovery of the Lymphatic System in Birds, Fish and Animals called Amphibians.* Cadell, London.
5. Christison, R. (1830). *Edinb. Med Surg. J.*, **33**, 276.
6. Chevreul, M. E. (1816). *Ann. Chimie Phys.*, **2**, 339.
7. Jensen, J. (1967). *Clio. Medica.*, **2**, 289.
8. Macheboeuf, M. and Rebeyrotte, P. (1949). *Discussions of the Faraday Society* **6**, 62.
9. Gofman, J. W., De Lalla, O. Glazier, F., Freeman, N. K., Nicholas, A. V., Strisower, E. H., and Tamplin, A. R. (1954). *Plasma*, **2**, 413.
10. Durrington, P. N. (1989). *Hyperlipidaemia. Diagnosis and Management.* Wright, London.
11. Mackness, M. I. (1989). *Biochem. Pharmacol.*, **38**, 385.
12. Steinberg, D., Parthasarathy, S., Carew, T. E., Khoo, J. C. and Witztum, J. L. (1989). *New Eng. J. Med.*, **320**, 915.
13. Durrington, P. N. (1990). *Scand. J. Clin. Lab. Invest.*, **50** (Suppl 198), 86.
14. Durrington, P. N., Bolton, C. H., and Hartog, M. (1978). *Clin. Chim. Acta*, **82**, 151.
15. Burstein M. and Legmann, P. (1982). In *Monographs on Atherosclerosis II* (ed. T. B. Clarkson and O. J. Pollak), S. Karger, Basle.
16. Burstein, M. and Samaille, J. (1960). *Clin. Chim. Acta*, **5**, 609.
17. Durrington, P. N. (1982). *CRC Crit. Rev. Clin. Lab. Sci.*, **18**, 31.
18. Bachorik, P. S. and Albers, J. J. (1986). In *Methods in Enzymology* (ed. J. J. Albers and J. P. Segrest), Vol. 129, p. 78. Academic Press, London.
19. Warnick, G. R. and Albers, J. J. (1978). *J. Lipid Res.*, **19**, 65.
20. Albers, J. J., Warnick, G. R., Johnson, N., Bachorik, P. S., Mussing, R., Lippel, K., and Williams, O. D. (1980). Lipid Res. Clin. Program Prevalence Study. *Circulation.* (Suppl IV), 9.
21. Warnick, G. R. and Albers, J. J. (1978). *Clin. Chem.*, **24**, 900.
22. Steele, B. W., Koehler, D. F., Azar, M. M., Blaszkowski, T. P., Kuba, K., and Dempsey, M. E. (1976). *Clin. Chem.*, **22**, 98.
23. Gries, A., Mimpf, J., Mimpf, M., Wurm, H., and Kostner, G. M. (1978). *Clin. Chim. Acta*, **164**, 93.
24. Friedewald, W. T., Levy, R. I., and Fredrickson, D. S. (1972). *Clin. Chem.* **18**, 499.

25. Winocour, P. H., Ishola, M., and Durrington, P. N. (1989). *Clin. Chim. Acta*, **179**, 79.
26. Wallentin, L. and Fahraena, L. (1981). *Clin. Chim. Acta*, **116**, 199.
27. Durrington, P. N. (1982). *Clin. Chim. Acta*, **120**, 21.
28. Skipski, V. P. (1972). In *Blood Lipids and Lipoproteins: Quantitation, Composition and Metabolism* (ed. G. J. Nelson) p. 471. Wiley-Interscience, New York.
29. Gidez, L. I., Miller, G. J., Burstein, M., Slagle, S., and Eder, H. A. (1982). *J. Lipid Res.*, **23**, 1206.
30. Kostner, M., Molinari, G., and Pilcher, P. (1985). *Clin. Chim. Acta*, **148**, 139.
31. Eyre, J., Hammett, F., and Miller, N. E. (1981). *Clin. Chim. Acta*, **114**, 225.
32. Hatch,' F. T. and Lees, R. S. (1968). *Adv. Lipid Res.*, **6**, 1.
33. Nichols, A. V., Kraus, R. M., and Mesliner, T. A. (1986). In *Methods in Enzymology* (ed. J. P. Segrest and J. J. Albers), Vol. 128, p. 417. Academic Press, London.
34. Towbin, H., Staehlin, T., and Gordon, J. (1979). *Proc. Natl. Acad. Sci. USA*, **76**, 4350.
35. Kane, J. W. and Gowland, E. (1986). *Ann. Clin. Biochem.*, **23**, 509.
36. Kahlon, T. S., Glines, L. A., and Lindgren, F. T. (1986). In *Methods in Enzymology* (ed. J. J. Albers and J. P. Segrest), Vol. 129, p. 26. Acdemic Press, London.
37. Patsch, J. R., Sailer, S., Kostner, G., Sandhofer, F., Holasek, A., and Brunsteiner, H. (1974). *J. Lipid Res.*, **15**, 356.
38. Shepherd, J., Packard, C. J., Stewart, J. M. Vallance, B. D., Lawrie, T. D. V., and Morgan H. G. (1980). *Clin. Chim. Acta*, **101**, 57.
39. Schumaker, V. N. and Puppione, D. L. (1986). In *Methods in Enzymology*, (ed. J. P. Segrest and J. J. Albers), Vol. 128, p. 155. Academic Press, London.
40. Havel, R. J., Eder, H. A., and Bragden, J. H. (1955). *J. Clin. Invest.*, **34**, 1345.
41. Carlson, K. (1973). *J. Clin. Path.*, **26** (Suppl 5), 32.
42. Lindgren, F. T., Jensen, L. C., and Hatch, F. T. (1972). In *Blood Lipids and Lipoproteins*. (ed. G. J. Nelson) p. 181. Wiley Interscience, New York.
43. Holmquist, L. (1982). *J. Lipid Res.*, **23**, 1249.
44. Chung, B. H., Segrest, J. P., Ray, M. J., Brunzell, J. D., Hokanson, J. E., Kraus, R. M., Beaudrie, K., and Cone, J. T. (1986). In *Methods In Enzymology* (ed. J. P Segrest and J. J. Albers), Vol. 128, p. 181. Academic Press, London.
45. Zak, B. (1977). *Clin. Chem.*, **23**, 1201.
46. Christie, W. W. (1987). *High performance lipid chromatography and lipids*. Pergamon Press, Oxford.
47. Warwick, G. R. (1986). In *Methods in Enzymology* (ed. J. J. Albers and J. P. Segrest), Vol. 129, p. 101. Academic Press, London.
48. Naito, H. K. and David, J. A. (1984). In *Lipid Research Methodology* (ed. J. A. Storey), Alan R. Liss Inc, New York 1.
49. Trinder, P. and Webster, D. (1984). *Ann. Clin. Biochem*, **21**, 430.
50. Cooper, G. R., Myers, G. L., Smith, S. J., and Sampson, E. J. (1988). *Clin. Chem.*, **34**, B95.
51. Naito, H. K. (1988). *Clin. Chem.*, **34**, B84.
52. Broughton, P. M. G. and Buckley, B. M. (1985). *Ann. Clin. Biochem.*, **22**, 547.
53. Dawber, T. R. (1980). *The Framingham Study. The Epidemiology of Atherosclerotic Disease*. Harvard University Press, Cambridge, Massachusetts.

54. Martin, M. J., Hulley, S. B., Browner, W. S., Kuller, L. H., and Wentworth, D. (1986). *Lancet*, **ii**, 933.
55. European Atherosclerosis Society Study Group (1987). *Europ. Heart J.*, **8**, 77.
56. Editorial 1989. *Lancet*, **i**, 1423.
57. Report of the National Cholesterol Education Program Expert Panel on Detection, Evaluation and Treatment of High Blood Cholesterol in Adults (1988). *Arch. Intern. Med.*, **148**, 36.
58. Lipid Research Clinics Program (1984). *J. Am. Med. Ass.*, **251**, 351.
59. Frick, M. H., Elo, O., Haap, K., Heinonen, O. P., Heinsalmi, P. Helo, P., Huttunen, J. K., Kaitaniemi, P., Koskinen, P., Manninen, V., Maenpaa, H., Malkonen, M., Manttari, M., Norola, S., Pasternack, A., Pikkarainen, J., Romo, M., Sjoblom, T., and Mikkila, E. A. (1987). *N. Engl. J. Med.*, **317**, 1237.
60. Blankenhorn, D. H., Nessim, S. A., Johnson, R. L., Sanmarco, M. E., Azen, S. P., and Cashin-Hamphill, L. (1987). *J. Am. Med. Ass.*, **257**, 3233.
61. Rifkind, B. M. and Segal, P. (1983). *J. Am. Med. Ass.*, **250**, 1869.
62. Lewis, B., Chait, A., Wootton, I. D. P., Oakley, C. M., Wrinkler, D. M., Sigurdsson, G., February, A., Mawer, B., and Birkhead, J. (1974). *Lancet*, **i**, 141.
63. Slack, J., Noble, N., Meade, T. W., and North, W. R. S. (1977). *Br. Med. J.*, **ii**, 353.
64. Thelle, D. S., Shaper, A. G., Whitehead, T. P., Bullock, D. G., Ashby, D., and Patel, I. (1983). *Br. Heart J.*, **49**, 205.
65. Mann, J. I., Lewis, B., Shepherd, J., Winder, A. F., Fenster, S., Rose, L., and Morgan, G. (1988). *Br. Med. J.*, **296**, 1702.
66. Armstrong, V. W., Cremer, P., Eberle, E., Manke, A., Schulze, F., Wieland, H., Kreuzer, H., and Seidel, D. (1986). *Atherosclerosis*, **62**, 249.
67. Durrington, P. N., Hunt, L., Ishola, M., Aroll, S., and Bhatnagar, D. (1988). *Lancet*, **i**, 1070.
68. Fredrickson, D. S., Levy, R. I., and Lees, R. S. (1967). *N. Engl. J. Med.*, **276**, 34, 94, 148, 215, and 273.
69. Beaumont, J. L., Carlson, L. A., Cooper, G. R., Fejfar, Z., Fredrickson, D. S., and Strasser, T. (1970). *Bull. Wld. Hlth. Org.*, **43**, 891.
70. Nikkila, E. A. (1983). In *The Metabolic Basis of Inherited Disease* (ed. J. B. Stanbury, J. B. Wyngaarden, D. S. Fredrickson, J. L. Goldstein, and M. S. Brown) 5th edn, p. 622. McGraw–Hill, New York.
71. Durrington, P. N. (1990). *Brit. Med. Bull.*, **46**, 1005.
72. Brown, M. S., Goldstein, J. L., and Fredrickson, D. S. (1983). In *The Metabolic Basis of Inherited Disease.* (ed. J. B. Stanbury, J. B. Wyngaarden, D. S. Fredrickson, J. L. Goldstein, and M. S. Brown) 5th edn, p. 672. McGraw–Hill, New York.
73. Brown, M. S., Goldstein, J. L., and Fredrickson, D. S. (1983). In *The Metabolic Basis of Inherited Disease.* (ed. J. B. Stanbury, J. B. Wyngaarden, D. S. Fredrickson, J. L. Goldstein, and M. S. Brown) 5th edn, p. 655. McGraw–Hill, New York.
74. Russell, D. W., Esser, V., and Hobbs, H. H. (1989). *Arteriosclerosis*, **9** (Suppl 1), 1.
75. Talmud, P., Tybjoerg-Hansen, A., Bhatnagar, D., Mbewu, A., Miller, J. P., Durrington, P. N., and Humphries, S. (1991). *Atherosclerosis*, **89**. 137.
76. Mahley, R. W. and Angelin, B. (1984). *Adv. Int. Med.*, **29**, 385.

77. Zannis, V. I. (1986). In *Methods in Enzymology* (ed. J. P. Segrest and J. J. Albers), Vol. 12, p. 823. Academic Press, London.
78. Fredrickson, D. S., Morganroth, J., and Levy, R. I. (1975). *Ann. Intern. Med.*, **82**, 150.
79. Herbert, P. N., Assman, G., Gotto, A. M., and Fredrickson, D. S. (1983). In *The Metabolic Basis of Inherited Disease*. (ed. J. B. Stanbury, J. B. Wyngaarden, D. S. Fredrickson, J. L. Goldstein, and M. S. Brown) 5th edn, p. 589. McGraw–Hill, New York.

2

Qualitative and quantitative separation of human apolipoproteins

HANS DIEPLINGER, HANS-JÜRGEN MENZEL, and
GERD UTERMANN

1. Introduction

Lipoproteins are the predominant lipid transport vehicles in human plasma. Their protein components are called apolipoproteins and represent a class of evolutionarily related polypeptides. The amino acid and nucleotide sequences as well as the chromosomal localization is known for all major apolipoproteins. *Table 1* shows the molecular weights and plasma concentrations of the different apolipoproteins. The various lipoproteins in human plasma have a characteristic apoprotein composition (for a review see reference 1). The protein moiety of high density lipoproteins (HDL) consists primarily of apo AI and apo AII with apolipoproteins apo AIV, C, D, and E as minor constituents. Within high density lipoproteins several subclasses exist that contain various combinations of these apolipoproteins (2). Low density lipoproteins are more homogeneous and their protein moiety contains

Table 1. Human apolipoproteins

Apolipoprotein	Mol. wt (× 10³)	Plasma conc. (mg/dl)	Purification methods (ref.)	Polymorphism (ref.)
AI	28	100−150	(17)	(13)
AII	17	30−50	(17)	(13)
AIV	46	15	(22)	(13)
B100	550	80−100	(30)	
B48	265	?	(31)	
apo(a)	300−800	> 0.1−> 30	(32)	(11)
CI	6.5	8	(25)	
CII	8.8	3−8	(26)	
CIII	8.9	8−15	(27)	
E	33	3−8	(23)	(7)

(exclusively) apo B100. Very low density lipoproteins also contain apo B100, but in addition apo E and the C-class apoproteins. VLDL subclasses with and without apo E have been described (3). The protein moiety of chylomicrons includes apo B48, the C-apoproteins, apo E and Apo AIV. All the above lipoprotein species are metabolically interrelated and exchange lipid as well as protein components in the course of their metabolic transformations in plasma. This is in contrast to the Lp(a)-lipoprotein which seems to be metabolically unrelated to the other plasma lipoproteins. Lp(a) consists of an LDL-like particle to which apolipoprotein(a) (apo(a)) is attached. The latter protein is responsible for the unique properties of Lp(a). Lp(a) has recently gained much interest because of the positive association of Lp(a) plasma levels with atherosclerotic diseases (for a review see reference 4). A high protein sequence homology of apo(a) with the plasma zymogen plasminogen has been demonstrated (5). Several studies have provided evidence that Lp(a) and apo(a) may interfere with fibrinolysis *in vitro* and it has been speculated that Lp(a) might be a link between atherogenesis and thrombo-genesis (6).

A large variety of methods for the qualitative and quantitative separation of apolipoproteins has been developed over the last decades.

For the qualitative analysis, various forms of electrophoretic methods have been used. The most important method certainly is the polyacrylamide gel electrophoresis (PAGE) in different variations. Since all apolipoproteins are hydrophobic, PAGE in the presence of SDS, sodium dodecyl sulphate, (SDS–PAGE) is especially useful with these proteins. Depending on the size of the apolipoproteins of interest, different concentrations of polyacrylamide (from 4–17%) as well as continous gradients of polyacrylamide concentrations are used. The second most important analytical technique for the character-ization of apolipoproteins is isoelectric focusing (IEF). Due to the already mentioned high hydrophobicity of apolipoproteins, IEF in gels containing urea has been especially useful. The combination of both IEF and SDS–PAGE has been shown to be a powerful high-resolution tool for the analytical separation of apolipoproteins. When these electrophoretic techniques are combined with immunoblotting, individual proteins can be detected by immunoreactions with the respective specific antibodies.

An even larger variety of techniques exists for the preparative separation of apolipoproteins. The classical preparative ultracentrifugation is still widely used (at least for crude preparations of lipoproteins). Frequently, ultra-centrifugation is the first step of a preparation scheme, followed by other methods of obtaining lipoproteins or apolipoproteins of high purity. These methods include chromatographic separations on various columns (anion- and cation-exchange-, chromatofocusing-, gel filtration-, hydroxylapatite-, immuno-affinity-, and hydrophobic interaction-chromatography) as well as preparative SDS–PAGE and preparative IEF.

2. Qualitative characterization of apolipoproteins

2.1 Methods for the qualitative separation of apolipoproteins from human plasma or apolipoprotein mixtures

2.1.1 Principle

Samples (either plasma or purified apolipoproteins) are subjected to SDS–PAGE, IEF, or two-dimensional electrophoresis. In the case of purified apolipoprotein, these techniques are perfectly suitable for the characterization of these proteins with respect to their identity, purity, and stability. For analysis of plasma or plasma fractions the proteins are transferred after electrophoresis by immunoblotting to nitrocellulose. The detection of the specific protein bands is achieved by immunological methods.

2.1.2 Specimen

Blood is collected into EDTA-containing tubes at a final EDTA concentration of 1.0 mg/dl. Plasma is obtained by immediate low speed centrifugation. 1.0 mM phenylmethylsulphonylfluoride (PMSF, Sigma), 0.4 μM aprotinin (Sigma), and 0.1 mM butylated hydroxy toluene (BHT, Sigma) are then added as protease inhibitors and antioxidant, respectively. The sample can then be stored at −20 °C and may be kept for months.

2.1.3 Reagents

Acrylamide, bis-acrylamide, ammonium persulphate, and TEMED (all ultrograde) are from Pharmacia-LKB. SDS (analytical grade) is from BDH Chemicals. Ampholyte in the pH range 4–6 and Blind Silane is from Pharmacia-LKB. Tris, boric acid, glycine, methanol, ethanol, ether, sodium deoxycholate, and urea are analytical grade from Merck. Urea solution is deionized with Serdolit MB (Serva) before use. Nitrocellulose 0.45 μm is from Schleicher and Schüll and AuCl$_3$, 4-chloro-1-naphthol and β-mercaptoethanol are from Sigma. Bovine serum albumin is from Boehringer. Sodium decylsulphate is from Kodak. *N*-ethylmorpholine, Triton X-100, and merthiolate is from Serva. The peroxidase- conjugated anti mouse IgG is from Dako and the anti rabbit IgG from Bio Makor. The antibodies against apolipoprotein E, AIV, and apo(a) were raised in our laboratory (7–9).

2.1.4 SDS–Polyacrylamide gel electrophoresis

i. Sample preparation

A 20 μl sample (containing 10 μg of protein), 5 μl of 20% SDS, 0.5 μl of mercaptoethanol, and 1 μl of 0.015% bromophenol blue in glycerol are mixed and boiled for 10 min; 20 μl are applied to the SDS gel.

ii. *Method for SDS electrophoresis of apolipoproteins*

SDS gels are prepared and electrophoresis is performed according to Neville (10) as described in *Protocol 1*. The gel system consists of a lower separating and a upper stacking gel. We use the Biometra electrophoresis chamber 'Minigel Twin'. The glass plates are treated before gel casting exclusively at the position of the upper gel with Bind Silane (15 µl Bind Silane in 20 ml ethanol and 5 ml 10% acetic acid). No trace of Bind Silane should reach the lower part of the glass plate.

Protocol 1. SDS–polyacrylamide gel electrophoresis of apolipoproteins

Reagents

(a) Lower 6.6% polyacrylamide gel
- i. Dissolve 20.68 g of Tris in H_2O, add concentrated HCl to give a pH of 9.18, make up the volume to 100 ml with water.
- ii. Dissolve 10.0 g of acrylamide and 0.1 g of bis-acrylamide in water to give a final volume of 25 ml.
- iii. Mix 120 µl of TEMED with 20 ml water.
- iv. Dissolve 50 mg ammonium persulphate in 25 ml H_2O.

(b) Upper 3.6% polyacrylamide gel
- i. Dissolve 2.62 g of Tris in water, add concentrated H_2SO_4 to adjust the pH to 6.14, and make up the volume to 100 ml with water.
- ii. Dissolve 10 g of acrylamide and 1.35 g of bis-acrylamide in water to give a final volume of 25 ml.

Procedure

1. Mix 4.2 ml of the Tris–HCl solution, 4.2 ml of the ammonium persulphate solution, 4.2 ml of the TEMED solution, and 1.4 ml of water.

2. Pour the mixture between two pretreated glass plates (9 × 9 cm with 1 mm spacers) until the gel reaches a height of 7 cm. Add a few drops of water-saturated *n*-butanol to obtain a straight surface. After polymerization the *n*-butanol is washed away with water.

3. Mix 2.5 ml of the Tris–H_2SO_4 solution, 0.9 ml of the acrylamide solution, 6.5 ml of the ammonium persulphate solution, and 12.5 µl of the TEMED solution.

4. Pour on top of the lower gel.

5. Insert the comb and allow polymerization of the gel to occur.

6. Remove the comb and apply the samples.

7. The lower tank buffer consists of Tris–HCl, pH 9.5: dissolve 51.7 g Tris in H_2O, adjust the pH to 9.5 by addition of 2 M HCl, and make up to 1 litre with H_2O.

8. The upper tank buffer consists of Tris–boric acid: dissolve 2.47 g boric acid, 4.92 g Tris, and 1.0 g SDS in H_2O, add solid Tris until a pH of 8.64 is reached and make up with H_2O to 1 litre.

9. Run the electrophoresis at 10 mA/plate until the bromophenol blue band is focused and then increase the current to 20 mA/plate with maximal 200 V for 1 to 2 h.

The described electrophoresis in 6.6% SDS gels is suitable for the separation of high molecular weight proteins and is also the first step required for phenotyping of apo(a) (11). For the separation and characterization of smaller proteins (MW < 100 kD) SDS gel electrophoresis in 10% poly-acrylamide gels has to be performed. For this purpose the protocol for preparing the lower polyacrylamide gel has simply to be changed from 2.8 ml to 4.2 ml acrylamide solution with no addition of H_2O.

2.1.5 Isoelectric focusing

i. Sample preparation

Samples containing up to 50 μg of protein are incubated in a total volume of 50 μl of 0.01 M Tris–HCl, pH 8.2, containing 1% decyl sulphate, 2% Ampholine (pH 4–6), 5 μl of β-mercaptoethanol, and 10 μl 80% sucrose for 1 h at room temperature and then directly applied into the electrophoresis slots.

Protocol 2. Isoelectric focusing of apolipoproteins
(according to Pagnan *et al.* (12) with modifications (13))

Reagents

(a) Dissolve 30 g acrylamide, 0.8 g bis-acrylamide, and 36 g urea in H_2O at a final volume of 100 ml.

(b) Dissolve 1 ml TEMED and 36 g urea in H_2O at a final volume of 100 ml.

Procedure

1. To 7 ml of the acrylamide solution, add 1.2 ml of Ampholines, 1.7 ml of the TEMED solution, 16 ml of 8 M urea, and 40 mg of ammonium persulphate.

2. Mix and pour between two glass plates size 14 × 12 cm (spacers 1.5 mm thick).

3. Insert an application comb and allow the gel to polymerize.

4. Remove the comb and apply the samples. They are protected against the upper tank buffer by a layering solution consisting of 1 ml 80% sucrose and 0.5 ml Ampholine in 8.5 ml H_2O.

Protocol 2. *Continued*

5. Assemble the glass plates in a SE 600 slab gel system (Hoefer, San Francisco, USA). Place 0.02 M NaOH in the upper tank and 0.01 M H_3PO_4 in the lower tank.

6. Focusing is run overnight at 250 V and 8 °C, starting with power limited to 3 W/plate, and continued at 800 V for 1 h in the morning.

2.1.6 Two-dimensional electrophoresis (13)

For 2-D electrophoresis, samples are first run on an IEF as in *Protocol 2*. Respective strips from the focusing gels are then cut out and immersed in 0.002 M ethylmorpholine–HCl, pH 8.5, 0.2% SDS, 0.1% β-mercaptoethanol, bromophenol blue solution, and 4% sucrose for 15 min at room temperature and applied to an SDS gel for electrophoresis in the second dimension. This SDS–PAGE is again performed according to Neville (10). After application and polymerization of the lower gel (usually containing 15% acrylamide) the butanol is removed and the gel surface is washed with water, layered with lower gel buffer, and stored overnight at 4 °C. Then the upper gel solution (5% acrylamide) is mixed and poured onto the lower gel to a height of 1.5 cm. After applying the gel strips from the IEF and a molecular weight standard, the electrophoresis apparatus is assembled and electrophoresis is carried out as described in *Protocol 1*.

2.1.7 Immunoblotting (Western blotting)

The transfer of proteins to nitrocellulose is achieved by the method of Towbin (14) with minor modifications.

Protocol 3. Procedure for immunoblotting onto nitrocellulose

1. Place the polyacrylamide gels under blotting buffer onto nitrocellulose (BA 85, 0.45 µm).

2. Place wet 3 mm thick filter paper on both sides and assemble with the cassettes of the blotting chamber (model TE42, Hoefer).

3. Use 192 mM glycine, 25 mM Tris, and 20% methanol in H_2O as blotting buffer.

4. Perform blotting of apo E and AIV for 1 h at 180 V and in the case of apo(a) overnight at 80 V. Keep the temperature at 10 °C.

Alternatively, immunoblotting can also be performed using the semi-dry blotting technique in a TE 70 SemiPhor apparatus from Hoefer Scientific Instruments (15). For the semidry blotting technique given in *Protocol 4*, a buffer consisting of 5.82 g Tris, 2.93 g Glycine, 1.9 ml of 20% SDS, 200 ml Methanol in 1 litre of H_2O is used.

Protocol 4. Alternative procedure for immunoblotting using a semi-dry technique

1. Immediately after electrophoresis soak the gels in blotting buffer for 15 min to equilibrate them.
2. Soak five pieces of filter paper and one piece of nitrocellulose membrane of exactly the size of the gel in the same buffer.
3. Place three pieces of buffer-saturated filter paper on top of the anode of the TE 70, followed by the buffer-soaked nitrocellulose sheet, the gel, and another two pieces of filter paper. During the build-up of such a sandwich, formation of air bubbles between the layers should be avoided to guarantee optimal electrotransfer.
4. Place the cover lid (cathode) on top of the stack and weight it down with a 1 kg weight.
5. Run the transfer at 250 mA and 15 V. Depending on the size of the protein, transfer times (at room temperature) are 30 min for proteins of less than 80 kD and 45 min for apo B and apo(a).

2.1.8 Immunodetection

Protocol 5. Procedure for immunodetection of apolipoproteins

Reagents

- Buffer A: 10 mM Tris–HCl, pH 9.0, 0.15 M NaCl, 0.01% merthiolate, 0.02% NaN_3
- Buffer B: Buffer A plus 0.1% SDS, 0.2% Triton X-100, 0.25% sodium deoxycholate
- Buffer C: Buffer A plus 1% bovine serum albumin
- TTBS: Dissolve 1.25 g Tween 20 in 2.5 litres TBS.
- TBS: Dissolve 50 ml 1 M Tris–HCl, pH 7.4 and 73 g NaCl in H_O to an end volume of 2.5 litres.
- Peroxidase substrate solution: 60 mg 4-chloro-1-naphthol, 20 ml methanol, 100 ml TBS, and 60 µl H_2O_2.

Procedure

1. After blotting, block the nitrocellulose for 0.5 h at 37 °C in buffer C and then immerse in buffer C plus the specific polyclonal rabbit antibodies to

Protocol 5. *Continued*

apo AI, apo E, AIV or mouse monoclonal antibodies to apo(a). These antibodies are either affinity purified polyclonal or monoclonal antibodies. The dilution range varies between 1:100 to 1:1000 depending on the specific concentration of the antibodies.

2. Incubate the nitrocellulose sheets for 2 h at room temperature, wash twice for a short period with buffer A, twice for 5 min in buffer B, and again two times in buffer A.

3. Subsequently submerge the sheets in buffer C plus the second antibody (peroxidase labelled anti mouse IgG (1:1000) for the monoclonal antibodies or gold-labelled anti rabbit IgG (1:150) for the polyclonal antibodies) and incubate for 2 h at room temperature. the gold-labelled antibodies are immediately visible whereas the peroxidase staining has to be developed. For this latter purpose the nitrocellulose is washed twice in TTBS (Tris-buffered saline + Tween) for 10 min.

4. Finally incubate the nitrocellulose sheets in peroxidase substrate solution for 10–40 min until a deep dark violet-black colour has developed.

5. Thereafter the nitrocellulose is washed with water extensively and can be stored in the dark.

2.1.9 Gold labelling of antibodies

The labelling of antibodies (*Protocol 6*) is performed according to Lin *et al.* (16). For this procedure a gold sol has to be prepared. For this purpose 40 mg $AuCl_3$ are dissolved in 400 ml redistilled H_2O and boiled under reflux for 0.5 h. The water and the glass ware (Repel-Silane treated) has to be completely free of traces of other metals. Then 10 ml of 0.1 M sodium citrate are added and the mixture is boiled for another 10 min. The gold sol can be stored for several months.

Protocol 6. Procedure for gold labelling of antibodies

1. Mix 10 ml of the gold sol with 200 µl 0.1 M polyethylene glycol (15 000–20 000 mol. wt.) and subsequently titrate to pH 7.6 with 0.1 M K_2CO_3. Measure the exact amount needed ($= x$).

2. Dilute the antibody solution (4 mg/ml) 1:10 with redistilled H_2O. Then dilute this stock solution into a serial dilution ranging from 1:2 to 1:32 with a volume of 1 ml.

3. To 100 µl of these dilutions add 1 ml of gold sol plus K_2CO_3 to give a pH of 7.6.

4. After 10 min, add 100 µl of 10% NaCl observing the change of colour from red to violet. The most diluted antibody concentration (1:y) that still shows red colour gives the right mixture ratio for the coupling of Au sol to the antibody solution.

5. For the final gold labelling, mix 150 ml gold sol, 15 times x ml K_2CO_3 to give a pH of 7.6, 15 ml anti rabbit IgG (1:10)$\times$(1:y), and 3 ml 0.1 M polyethylene glycol.

6. Centrifuge the mixture for 2 h at 10 000 r.p.m. at 4 °C in glass tubes. Dissolve the pelleted Au-marked antibody in 3 ml of 0.1 M potassium phosphate pH 7.4, containing 4% polyvinylpyrrolidone, 0.2 mg/ml 0.1 M polyethylene glycol (15–20 000), and 0.05% NaN$_3$.

7. Centrifuge this solution for 0.5 min at 10 000 r.p.m. and discard the pellet. The gold-labelled antibody is ready for immunostaining and used in a 1:200 dilution.

2.2 Detection of mutants and polymorphisms of apolipoproteins

All common mutations and polymorphisms of human apolipoproteins AI, AII, AIV, and E are detected by isoelectric focusing of native or delipidated total human plasma (7, 13). The size polymorphism of apo(a) is determined by SDS–PAGE of reduced human plasma (11). After electrophoresis the proteins are transferred by immunoblotting to nitrocellulose. The detection of the specific protein bands is achieved by immunological methods.

2.2.1 Preparation of samples

i. Apo AI, AII, AIV (13)

Incubate 4 µl plasma with 50 µl of 0.01 M Tris–HCl, pH 8.2, containing 1% sodium decyl sulphate, 2% Ampholine (pH 4–6), 5 µl of β-mercaptoethanol, and 10 µl 80% sucrose for 1 h at room temperature and then apply directly into the electrophoresis slots.

ii. Apo E (7)

Delipidate 10 µl plasma with 2.5 ml ethanol:ether (3:1) overnight at −20 °C as described in Chapter 4. **Extreme precautions must be taken to avoid explosion caused by the low flash point of ether.** Centrifuge the precipitate at 5000 r.p.m. for 10 min at −10 °C and wash once with ether for 1 h at −10 °C. After centrifugation dissolve the precipitate in 200 µl of 0.1 M Tris–HCl, pH 10.0, 6 M urea, 1% sodium decyl sulphate, and 2 µl of β-mercaptoethanol and leave for 0.5 h at 4%°C. Apply 20 µl onto the isoelectric focusing gel.

iii. Apo(a) (11)

Boil 2 µl plasma in 50 µl 5% SDS, 0.02 M N-ethylmorpholine–HCl, pH 8.5, 2 µl β-mercaptoethanol, and 4 µl of 0.015% bromophenol blue in glycerol. Apply 5 µl of this mixture to the 6.6% SDS polyacrylamide gel (see *Protocol 1*).

3. Quantitative separation of apolipoproteins

3.1 Isolation of apo AI and apo AII

Human plasma apo AI and AII are isolated from HDL which is obtained from plasma from fasting normal subjects by sequential preparative ultracentrifugation (according to reference 17, with modifications).

Protocol 7. Procedure for the quantitative isolation of apo AI and apo AII

1. Adjust the density of plasma to 1.063 g/ml with solid KBr and remove the VLDL and LDL fractions by ultracentrifugation at 50 000 r.p.m. for 24 h at 5 °C in a Beckman 50.2 Ti rotor.

2. Adjust the density of the infranate, (obtained by tube slicing) to 1.21 g/ml and centrifuge at 50 000 r.p.m. for 48 h at 5 °C in the same rotor.

3. The supernate from this centrifugation contains the HDL. Dialyse the supernate against 1 mM EDTA, pH 7.4.

4. Bring the dialysed HDL to 6 M guanidine–HCl and incubate for 3 h at 37 °C.

5. After another dialysis against 1 mM EDTA, pH 7.4 adjust the density of the HDL to 1.21 g/ml and centrifuge at 50 000 r.p.m. for 24 h at 5 °C.

6. The floating top portion contains the apo AII, whereas the infranate contains the apo AI. Dialyse both fractions against 0.15 M NaCl, 1 mM EDTA, pH 7.4, and delipidated three times with ethanol/ether (3/1) at −20 °C.

7. Dissolve the delipidated samples in 30 mM Tris–HCl, pH 8.0, 6 M urea and apply to an anion exchange chromatography column (Mono Q HR 10/10, Pharmacia). Elute with a linear 300 ml salt gradient (0–0.15 M NaCl, 30 mM Tris HCl, pH 8.0, 6 M urea) at a flow rate of 4 ml/min to obtain pure apo AI and apo AII.

8. Fractions containing the respective pure apolipoproteins (as checked by SDS–PAGE, see above) are pooled, dialysed against PBS, and stored at −20 °C (1 mg/dl).

Alternative methods to purify apolipoproteins AI and AII include hydroxylapatite chromatography from ultracentrifugally prepared HDL (18), affinity chromatography from whole plasma (2) as well as chromatofocusing

(19) or reversed phase chromatography (20) of the delipidated HDL fraction. All of these methods are either more tedious to perform or do not produce sufficient purities so that the relatively simple ion exchange chromatography is preferred.

3.2 Isolation of apo AIV

Human apolipoprotein AIV is purified from lipoprotein-depleted serum by absorption on a triglyceride–phospholipid emulsion (as described by Weinberg et al. (21) followed by anion-exchange chromatography of the delipidated lipid-bound proteins (as described by Steinmetz et al. (22), with modification).

Protocol 8. Procedure for the quantitative isolation of apo AIV

1. Prepare lipoprotein-depleted serum (LPDS) by ultracentrifugation of 500 ml serum of a density of 1.25 g/ml at 50 000 r.p.m. and remove the floating lipoproteins.

2. Dialyse the supernatant against 0.15 M NaCl, 50 mM potassium phosphate, pH 7.4, and 0.05% Na_2EDTA.

3. In the meantime prepare 400 ml Intralipid-emulsion (Kabi) by centrifuging 400 ml of Intralipid for 35 min at 25 000 r.p.m. and 4 °C in an SW 28 rotor (Beckman), removing the floating lipids and resuspending them with PBS to the original volume.

5. Incubate 400 ml of LPDS with 400 ml of PBS, 225 g NaCl, 160 ml Intralipid (prepared as described above) at pH of 7.4 (adjusted with KOH) in a shaking waterbath at 37 °C.

6. Reisolate the Intralipid by centrifugation in the SW 28 rotor at 27 000 r.p.m. for 20 min at 4 °C. Then resuspend the supernatant lipids in 10 ml PBS and dialyse against 50 mM Tris–HCl pH 8.2.

7. Delipidate the dialysed material by injecting it with a narrow-gauge needle directly into 1000 ml of ethanol/ether (3/1), incubate for 24 h at −20 °C, then centrifuge the pellet and resuspend.

8. Repeat this procedure with 800 ml for 24 h, 400 ml for 4 h, 80 ml for 24 h, and 80 ml for 3 h. The last delipidation step is performed with 30 ml of ether for 2 h. All incubations are performed at −20 °C.

9. Dissolve the delipidated pellet in 7.2 M urea, 10 mM Tris–HCl, pH 8.0, and evaporate the remaining ether by degassing with N_2. After centrifuging some undissolved material, the supernatant is then ready for further purification by anion exchange chromatography.

The FPLC (fast protein liquid chromatography) system (Pharmacia) incorporating a Mono Q HR 16/10 column is used to separate apo AIV from other intralipid-bound proteins. The system consists of 2 P-500 gradient pumps and a GP/-250 gradient programmer to form the gradient. The elution is monitored by a UV monitor at 280 nm and a chart recorder documenting UV absorbance, the programmed gradient and the collected fractions (2 ml fraction size). The whole chromatography is performed at room temperature. The column is equilibrated with 10 mM Tris–HCl, pH 8.0, 7.2 M urea at a flow rate of 1 ml/min and the apo AIV containing material applied. The column is then eluted with a linear 150 ml gradient of 0–300 mM NaCl at a flow rate of 1 ml/min. Eluted fractions are immediately dialysed to remove urea and analysed by 15% SDS–PAGE and Western Blotting. Three major peaks are eluted by this method of chromatography; the first containing mainly apo AI, the second containing pure apo IV, and the third containing mainly albumin.

3.3 Isolation of apo E

Apolipoprotein E is separated from apo VLDL by preparative SDS–PAGE (23).

Protocol 9. Procedure for the quantitative isolation of apo E

1. Separate VLDL from plasma by ultracentrifugation at its own density as described for the LDL separation (Section 3.5.1).
2. Obtain apo VLDL by delipidation with acetone/ethanol 1:1 (v/v) at − 20 °C overnight.
3. Perform preparative PAGE using the discontinuous buffer system of Neville (10), with an acrylamide gel concentration of 15%. Dissolve up to 50 mg of apoprotein in 2 ml of 0.04 M Tris–borate, pH 8.64, containing 2% SDS and 1% 2-mercaptoethanol, and boil for 3 min at 100 °C.
4. Add glycerol to the sample and apply to the vertical gel column. Apply a constant voltage of 50 V resulting in a current of about 8 mA.
5. Elute with 0.02 M ethylmorpholine/HCl buffer, pH 9.2 at a flow rate of 4.4 ml/h. After monitoring the flow at 280 nm, collect 2 ml fractions and pool the pure apo E fractions (checked by SDS–PAGE and immuno-blotting).

Alternatively, the preparative SDS–PAGE may be run on a vertical slab gel system (24). A 1-mm thick 5–20% acrylamide gradient gel is prepared. 15 mg of apo VLDL dissolved in 2 mol of 0.125 M Tris–HCL, pH 6.5, 2% (w/v) SDS, 10% (v/v) glycerol, and 3% (v/v) β-mercaptoethanol are boiled as described before and applied to the gel. A constant current of 50 mA is used until the voltage reaches 300 V. After the run the protein bands are visualized with KCl as described for the preparation of B48 (see Section 3.5). The apo

E-band is cut out and eluted from the gel with distilled water. Then it is dialysed against 10 mm Tris–HCl, pH 8.0, 0.05% SDS to remove the KCl, and lyophilized.

Apo E can also be purified by gel filtration chromatography of the delipidized VLDL in the presence of SDS (23). The separation is performed at room temperature on a 2.5 × 100 cm Sephadex G-200 column (Pharmacia) equilibrated with 0.01 M ethylmorpholine/HCl, pH 8.5, 0.1% SDS, 1 mM DTE.

3.4 Isolation of apolipoproteins C

All three species of apolipoproteins C are isolated by the same technique, namely starting with delipidated VLDL and performing a gel filtration chromatography (to remove the larger apolipoproteins B and E) followed by an anion exchange chromatography (25–27). Preparation of VLDL and its delipidation is performed as described for apo E.

The gel filtration chromatography is performed in 1 mM sodium decyl sulphate. The fractions containing the C-proteins are pooled and chromatographed on a DEAE anion exchange chromatography column in 8 M urea. The fractions containing the respective C-proteins are pooled and desalted by a PD-10 column. In the case of apo CII, the pool from the DEAE chromatography is lyophilized and rechromatographed on Sepharose G-75 in a buffer containing 0.01 M Tris–HCl, pH 8.6, 5.4 M urea (26).

Alternatively, the purification of apo CII and CIII has been recently described by preparative High Pressure Liquid Chromatography (HPLC) of the delipidated VLDL apolipoproteins (28). These proteins are solubilized in 0.01 M Tris–HCl, pH 8.2, containing 6 M urea and centrifuged for 15 min at 3000g to remove insoluble material (mainly apolipoprotein B). The supernatant containing the urea soluble VLDL apolipoproteins (approx. 3 mg) is applied on a reversed phase column (Nucleosil 1000 C_4, 250 × 4.6 mm i.d., 7 μm, Chrompack) of a Varian Model 540 HPLC apparatus. The separation is carried out at room temperature by gradient elution using 50 mM ammonium acetate, pH 6.0, as solvent A and 2-propanol as solvent B. A 60 ml linear gradient from 100% A to 50% A/50% B at a flow rate of 1 ml/min is used. Fractions containing apo CII and CIII are pooled, concentrated by a Speed Vac lyophilization step and rechromatographed as described above. Before injection, each sample is filtered through a 0.22 μm Millipore filter, type GS.

Another method for the quantitative separation of the C apoproteins has been developed using preparative size-exclusion (gel filtration) chromatography of delipidated HDL apolipoproteins (29). HDL are prepared by classical sequencial ultracentrifugation and delipidated with ethanol/ether. This HDL apolipoprotein mixture is then separated by preparative HPLC using a Varian Model 5060 chromatograph and a MicroPak TSK 3000SW preparative size exclusion column (22 × 300 mm).

3.5 Isolation of apo B100 and B48

Due to the high molecular weight and hydrophobicity of these two proteins it is impossible to purify them in lipid-free form without detergents. The following protocols therefore describe methods for purification of the corresponding lipoprotein classes (LDL and chylomicrons, respectively).

3.5 Isolation of LDL (pure Lp B)

LDL are purified by classical stepwise density ultracentrifugation followed by various techniques for removing protein and lipoprotein impurities (albumin, apo A class, apo C class, E, Lp(a)) which are always copurified even after recentrifugation of the LDL (30).

Protocol 10. Procedure for the separation of apo B

1. Obtain fresh plasma by low-speed centrifugation and supplement with gentamycin sulphate (0.1 mg/ml), chloramphenicol (0.05 mg/ml), and sodium-azide (0.2 mg/ml) to prevent microbial degradation of apo B. Add EDTA (1 mM) and PMSF (0.05 mg/ml) to prevent proteolytic degradation. All buffers and density solutions must contain these same additives.

2. Remove VLDL and chylomicrons from plasma by centrifugation at plasma density ($d = 1.006$ g/ml) for 18 h at 5 °C and 40 000 r.p.m. in a 50.2 Ti Beckman rotor. Then adjust the density of the infranatant to 1.063 g/ml by addition of solid KBr and spin for 24 h at 50 000 r.p.m. at 5 °C in the same rotor.

3. Obtain the floating material (containing the LDL) by aspiration or tube slicing.

4. Methods for further purification include preparative SDS–PAGE (described for apo B48), anion exchange chromatography, binding to artificial lipid emulsions, and immunaffinity chromatography.

For the anion exchange chromatography, the LDL from the ultracentrifugation is dialysed against 50 mM NaCl, 10 mM Tris–HCL, pH 7.4, and applied on a HiLoad 16/10 Q Sepharose column (Pharmacia) which was equilibrated with the same buffer. The elution is performed on an FPLC system using a 120 ml linear 50–500 mM NaCl gradient at room temperature at a flow rate of 1 ml/min. Under these conditions, the LDL is free from contaminating proteins as checked by silver-stained PHAST SDS–PAGE (Pharmacia).

Alternatively, LDL can be incubated with Intralipid (10%, Kabi) (1:1, v/v) for 1 h at 37 °C and centrifuged at 20 000g for 30 min (30). By this method, the apo A and C-class proteins can be removed because they have a high affinity for such emulsions. Other proteins are consecutively removed by immunoaffinity chromatography on columns to which affinity-purified poly-colonal antibodies to the respective proteins are covalently linked.

3.5.2 Isolation of apo B48

This protein can be obtained by preparation of chylomicrons, followed by delipidation and preparative SDS–PAGE (31). Chylomicrons are obtained from lymph by ultracentrifugation in a 50.2 Ti Rotor at 39 000 r.p.m. for 18 h at 4 °C. The top layer is then collected, resuspended in PBS, and the ultracentrifugation repeated. The chylomicrons are then delipidated with ethanol/ether (3:1, v/v) and apo B48 isolated by preparative 5% PAGE in the presence of SDS. The delipidated samples are prepared as for analytical PAGE. The buffer in the gel and in both tanks is 0.1 M Na_2HPO_4 (pH 7.2) with 0.1% SDS. The tank buffer also contains 0.1 mg/ml gentamycin sulphate and sodium azide (0.2 mg/ml). For visualization of the protein bands the gel is incubated overnight in 0.2 M KCl at 4 °C, which leads to precipitation of SDS in the gel.

3.5.3 Isolation of Lp(a)

Blood donors with plasma Lp(a) concentrations greater than 50 mg/dl are selected for obtaining whole plasma for isolation of Lp(a). To 600 ml of such a plasma the following antiproteolytic and anti-microbial agents are added: 0.26 ml PMSF (100 mg/ml ethanol), 120 mg NaN_3, 60 mg sodium ethylmercury thiosalycylate, 0.45 ml aprotinin (2 mg/ml), and 0.6 ml 0.5 M Na_2EDTA.

The purification scheme consists of a sequential ultracentrifugation procedure. If apo(a) instead of Lp(a) is needed, then the purified Lp(a) is dissociated into its apo B- and apo(a)-containing constituents by incubation with a disulphide-reducing agent followed by another ultracentrifugation step which removes the LDL moiety by flotation and leaves the apo(a) in the bottom fraction (32).

Protocol 11. Procedure for the isolation of Lp(a)

1. Ultracentrifuge the 480 ml plasma sample for 48 h at 20 °C and 49 000 r.p.m. in a 50.2 Ti Beckman rotor at a plasma density of 1.063 g/ml (adjusted with solid KBr).

2. Recover the Lp(a)-containing lipoprotein fraction by aspiration from the middle part of the tube between the floating VLDL and LDL and the HDL and other serum proteins of the bottom fraction.

3. Then adjust the Lp(a) fraction to density 1.10 g/ml and centrifuge for 24 h at 20 °C and 49 000 r.p.m. This step removes remaining HDL impurities

Protocol 11. *Continued*

which sediment at the bottom of the tube whereas the floating lipoprotein fraction contains Lp(a).

4. In order to obtain the apo(a) protein, incubate the purified Lp(a) from the second ultracentrifugation for 1 h at room temperature in the presence of 1 mM dithioerythritol (DTE). The incubation should be performed in stoppered glasstubes under an atmosphere of N_2.

5. After this incubation, centrifuge the Lp(a) preparation in a SW 41 Beckmann rotor for 3 h at 40 000 r.p.m. and 20 °C. For this purpose 3 ml of the Lp(a) preparation at the original density of 1.10 g/ml is underlayered under 3 ml of density solution 1.035 g/ml and 2ml of density solution of 1.019 g/ml. Finally, approximately 4 ml of density solution 1.21 g/ml are underlayered at the bottom of the tube.

6. After centrifugation, recover the bottom 3 ml of the tube containing apo(a) by downwards fractionation.

7. Finally dialyse the apo(a) preparation against 0.2 mM ethyl-morpholine, pH 8.5, and add the anti-proteolytic and anti-microbial agents.

Alternatively, the Lp(a) containing fraction from the first ultracentrifugation can be further purified by gel filtration chromatography on a Sepharose 4B column (33). 5 ml of lipoprotein solution containing 200–250 mg of protein are applied to the column (2.5 × 100 cm) after dialysis against the eluant buffer (0.1 M Tris–HCl, pH 9.0). 5 ml fractions are collected, assayed for protein at 280 nm and screened for Lp(a) by SDS–PAGE and immunoblotting.

A further alternative method is immunoaffinity chromatography on an anti-apo(a)-affinity column (32). The Lp(a) containing fraction from the first ultracentrifugation is directly applied to the column (best performed by overnight rotating of the stoppered column), washed with PBS and eluted with 0.2 M glycine–HCl, pH 2.0, 6 M urea, 0.15 M NaCl. The eluate is immediately neutralized with 1 M K_2HPO_4, pH 9.8, dialysed against PBS, and the anti-proteolytic and anti-microbial agents added.

Lysine–Sepharose affinity chromatography takes advantage of the lysine-binding properties of apo(a) (due to the homology with plasminogen) and therefore is another alternative method for Lp(a) purification (34). The Lp(a)-enriched fraction from the ultracentrifugation of plasma is dialysed against 0.1 M phosphate, pH 7.4, 1 mM EDTA, and applied to a 7 × 2 cm column of lysine–Sepharose 4B equilibrated with the same buffer. After washing the unbound proteins, the Lp(a) bound to the column is released by the same buffer containing 50 mM 6-aminohexanoic acid.

Acknowledgements

The excellent technical assistance of Linda Fineder, Eva-Maria Lobentanz, and Ulrike Scheidle is gratefully acknowledged. Parts of the work described here were supported by grants from the Austrian 'Fonds zur Förderung der Wissenschaftlichen Forschung' (Projects S 4604 and S4610) to H.D. and G.U.

References

1. Mahley, R. W., Innerarity, T. L., Rall, S. C., Jr., and Weisgraber, K. H. (1984). *J. Lipid Res.*, **25**, 1277–94.
2. Cheung, M. C. and Albers, J. J. (1984). *J. Biol. Chem.*, **259**, 12201–9.
3. Trezzi, E., Calvi, C., Roma, P., and Catapano, A. L. (1983). *J. Lipid Res.*, **24**, 790–5.
4. Utermann, G. (1989). *Science*, **246**, 904–10.
5. McLean, J. W., Tomlinson, J. E., Kuang, W.-J., Eaton, D. L., Chen, E. Y., Fless, G. M., Scanu, A. M., and Lawn, R. M. (1987). *Nature*, **300**, 132–7.
6. Loscalzo, J., Weinfeld, M., Fless, G. M., and Scanu, A. M. (1990). *Arteriosclerosis*, **10**, 240–5.
7. Menzel, H. J. and Utermann, G. (1986). *Electrophoresis*, **11**, 492–5.
8. Utermann, G. and Beisiegel, U. (1979). *Eur. J. Biochem.*, **99**, 333–43.
9. Kraft, H. G., Dieplinger, H., Hoye, E., and Utermann, G. (1988). *Arteriosclerosis*, **8**, 212–16.
10. Neville, D. M. (1971). *J. Biol. Chem.*, **246**, 6328–34.
11. Utermann, G., Menzel, H. J., Kraft, H. G., Duba, H. C., Kemmler, H. G., and Seitz, C. (1987). *J. Clin. Invest.*, **80**, 458–65.
12. Pagnan, A., Havel, R. J., Kane, J. R., and Kotite, L. (1977). *J. Lipid Res.*, **18**, 613–22.
13. Menzel, H. J., Kladetzky, R. G., and Assmann, G. (1982). *J. Lipid Res.*, **23**, 915–22.
14. Towbin, H., Staehlin, T., and Gordon, J. (1979). *Proc. Natl. Acad. Sci. USA*, **76**, 4350–4.
15. Kyhse-Anderson, J. (1984). *J. Biochem. Biophys. Meth.*, **10**, 203–9.
16. Lin, N. S. and Langenberg, W. G. (1983). *Ultrastruct. Res.*, **84**, 16–23.
17. Cheung, M. C. and Albers, J. J. (1977). *J. Clin. Invest.*, **60**, 43–50.
18. Kostner, G. M. and Holasek, A. (1977). *Biochim. Biophys. Acta*, **488**, 417–31.
19. Jauhiainen, M. S., Laitinen, M. V., Penttilä, I. M., and Puhakainen, E. V. (1982). *Clin. Chim. Acta*, **122**, 85–91.
20. Weinberg, R., Patton, C., and DaGue, B. (1988). *J. Lipid Res.*, **29**, 819–24.
21. Weinberg, R. B. and Scanu, A. M. (1983). *J. Lipid Res.*, **24**, 52–9.
22. Steinmetz, A., Clavey, V., Vu-Dac, N., Kaffamik, H., and Fruchart, J.-C. (1989). *J. Chromatogr.*, **487**, 154–60.
23. Utermann, G. (1975). *Hoppe-Seyler'S Z. Physiol. Chem.*, **356**, 1113–21.
24. Wardell, M. R., Brennan, S. O., Janus, E. D., Fraser, R., and Carrell, R. W. (1987). *J. Clin. Invest*, **80**, 483–90.

25. Shulman, R. S., Herbert, P. N., Wehrly, K., and Fredrickson, D. S. (1975). *J. Biol. Chem.*, **250**, 182–90.
26. Jackson, R. L., Baker, H. N., Gilliam, E. B., and Gotto, A. M. Jr (1977). *Proc. Natl. Acad. Sci. USA*, **74**, 1942–5.
27. Brewer, H. B., Jr, Shulman, R., Herbert, P. N., Ronan, R., and Wehrly, K. (1974). *J. Biol. Chem.*, **249**, 4975–84.
28. Beyne, P., Doute, M., Lacour, B., Lederer, F., and Pays, M. (1990). *J. Chromatogr. Biomed. Appl.*, **527**, 140–5.
29. Wehr, C. T., Cunico, R. L., Ott, G. S., & Shore, V. G. (1982). *Anal. Biochem.*, **125**, 386–94.
30. Zechner, R., Moser, R., and Kostner, G. M. (1986). *J. Lipid Res.*, **27**, 681–6.
31. Kane, J. P., Hardman, D. A., and Paulus, H. E. (1980). *Proc. Natl. Acad. Sci. USA*, **77**, 2465–9.
32. Menzel, H.-J., Dieplinger, H., Lackner, C., Hoppichler, F., Lloyd, J. K., Muller, D. R., Labeur, C., Talmud, P. J., and Utermann, G. (1990). *J. Biol. Chem.*, **265**, 981–6.
33. Ehnholm, C., Garoff, H., Simons, K., and Aro, H. (1971). *Biochim. Biophys. Acta*, **239**, 431–9.
34. Armstrong, V. W., Harrach, B., Robenek, H., Helmhold, M., Walli, A. K., and Seidel, D. (1990). *J. Lipid Res.*, **31**, 429–41.

Immunological methods for studying and quantifying lipoproteins and apolipoproteins

ROSS W. MILNE, PHILIP K. WEECH, and YVES L. MARCEL

1. Introduction

Antibodies and immunochemical techniques have been used to quantify lipoproteins and apolipoproteins and to study their structure, function, metabolism, heterogeneity, and genetic polymorphism. The antibodies used are, in general, directed against the apolipoproteins or against molecules that are implicated in lipoprotein metabolism such as the triglyceride lipases, cell surface lipoprotein receptors, lecithin:cholesterol acyltransferase, and lipid transfer proteins. Due to their complex structure, lipoproteins have a number of distinct antigenic properties that impose certain modifications to the usual criteria that are applied in the production, characterization, and use of antibodies. In this chapter we will emphasize the unique antigenic features of lipoproteins and how these peculiarities influence the methodologies that are used to detect and measure lipoproteins. The techniques that are described are those that we currently use in our laboratories. For other techniques and a description of the production and use of antibodies in general, the reader is referred to reference (1).

2. Apolipoproteins and lipoproteins as immunogens

The lipid environment of the apolipoprotein can have a major influence on its immunoreactivity. In extreme cases, antibodies raised against the delipidated apolipoprotein can fail to react with the antigen when it is presented in a native lipoprotein (2) and antibodies prepared against the native lipoprotein do not necessarily react with the delipidated apoliproproteins (3). More often, the modulation of immunoreactivity of the apolipoprotein by the physical and chemical properties of the lipoprotein particle is more subtle (4). The respective roles of particle composition, size, and density in this modulation are only beginning to be defined. Apolipoproteins can also undergo spontaneous *in vitro* chemical modifications during storage at 4 °C that can dramatically increase or decrease their immunoreactivity (5, 6). In many cases such modifications can be avoided by including protease inhibitors in the lipoprotein preparations and by taking measures to avoid oxidation (7). Freezing and lyophilization of lipoproteins can also alter their immunoreactivity.

The special antigenic properties of lipoproteins have practical implications for the production of antibodies and for the methods of quantitation of apolipoproteins. It is often preferable to use the native lipoprotein rather than the delipidated purified apolipoprotein as immunogen. When this is impractical (e.g. for the preparation of polyclonal antibodies), one may consider reincorporation of the purified apolipoprotein into lipid vesicles in the hope that the apolipoprotein will reassume a native conformation (3). During preparation of the immunogen, care should be taken to avoid chemical modification or degradation to minimize the possibility of generation of antibodies to modified variants. Similar precautions should be taken to limit chemical modification in samples in which apolipoproteins are to be quantified. An emulsion of antigen in Freund's Adjuvant is commonly used for immunization. The mineral oil in the adjuvant and the emulsification procedure could have a disruptive effect on lipoproteins and alter the conformation of apolipoproteins. Alternative adjuvants such as muramyl dipeptide (8) (Calbiochem Corp.) may therefore be preferable.

We have had little success in raising antibodies to synthetic peptides that mimic the primary structure of apolipoproteins. Almost without exception these anti-peptide antisera showed excellent reactivity with denatured forms of the respective apolipoproteins (e.g. on Western blots) but failed to recognize the same molecules in their native conformations.

3. Production of antibodies

3.1 General considerations

Several factors can enter into the decision whether to prepare monoclonal or polyclonal antibodies. Both approaches have their respective advantages and disadvantages. Preparation of polyclonal antisera is much less labour-intensive and does not require facilities for, or expertise in cell culture. On the other hand, purified immunogen is not required for the production of monoclonal antibodies (mAbs) and the supply of an antibody of constant affinity and specificity should be unlimited. The avidities of antisera are often higher than those of individual mAbs and can be used in assays that require direct immunoprecipitation of antigen (e.g. single radial immunodiffusion and rocket immunoelectrophoresis). Heterogeneity in expression of apolipoprotein epitopes may also present fewer problems in immunoassays using polyclonal antibodies. On the other hand, antibody specificity is much easier to achieve and validate with monclonal antibodies and, due to their unique intramolecular specificities, mAbs are superior probes of lipoprotein structure, heterogeneity, and function. It should be noted that in some cases it is possible to obtain certain characteristics of a polyclonal antiserum by mixing mAbs of different intramolecular specificities.

3.2 Preparation of polyclonal antisera

With the exception of LDL which contains virtually only apo B, the other lipoproteins contain various apolipoproteins species. Therefore, to obtain specific anti-apolipoprotein antisera, one is almost obliged to use delipidated and purified protein as the antigen. In the rabbit, 50 to 100 µg together with adjuvant is normally administered per injection. High affinity antisera can, however, be obtained using as little as 10 µg of antigen per injection. It should be noted that the risk of eliciting antibodies to other molecules that contaminate the antigen preparation increases when the total protein injected is increased. Rabbits are normally the animal of choice for raising polyclonal antibodies. The immunization and bleeding protocols that have been established for other antigens are equally applicable to lipoproteins and apolipoproteins (1).

3.3 Characterization of antisera

The degree of assurance that an antiserum is specific is a function of the sensitivity of the methods used to test the specificity. The antiserum must therefore be shown to be specific in the assay for which it is to be used. It is recommended that anti-apolipoprotein sera be tested for specificity in the antibody capture assay (*Protocol 2*), in a competition immunometric assay (*Protocol 6*) and by immunoblotting after one- or, preferably, two-dimensional polyacrylamide gel electrophoresis of lipoproteins and whole plasma (Section 4). Polyspecific antisera can often be rendered monospecific by immunodepletion of the contaminating antibodies with the appropriate antigen(s). As an example, an anti-apo E antiserum that shows some reactivity with apo AI can be immunoabsorbed with purified apo AI or apo E-free HDL. To avoid subsequent problems with soluble immune complexes, it is preferable to insolubilize the antigen used for the immunoadsorption (*Protocol 7*). Contaminating antibodies can be removed by either batch or column immunoadsorption.

Affinity of the antiserum can most easily be determined by a liquid phase competitive radioimmunoassay using as the competing antigen either the purified apolipoprotein or the intact lipoprotein (9). As the physical and chemical properties of lipoproteins can influence the immunoreactivity of apolipoproteins, the abilities of an antiserum to recognize its specific antigen when this latter is in different lipid environments should also be evaluated (e.g. in competitive radioimmunometric assay (*Protocol 6*) and by immuno-precipitation (Section 6.2)).

3.4 Preparation of mAbs

An advantage of the production of antibodies using hybridoma technology is that one is not limited to the use of purified antigen for immunization. This has particular implications for the production of antibodies against apolipo-proteins. For the reasons discussed above, it may be advantageous to immunize with native lipoprotein particles (which may contain several different apolipoproteins) rather than with delipidated, purified apolipo-proteins.

Both rats and mice can be used to produce mAbs. Based on the reported intramolecular specificities of rat mAbs, different epitopes on human apolipoproteins may be immunodominant in the rat compared to the mouse (10). The amounts of ascites that may be obtained from a hybridoma-bearing rat are proportionally greater than those that can be obtained from a mouse. A number of mouse and rat myeloma cell lines are available for use as fusion partners for the production of mAbs (1). Most are hypoxanthine-guanine phosphoribosyltransferase negative and will, therefore, be killed in medium supplemented with hypoxanthine, aminopterine, and thymidine (HAT) (11) and most are negative for endogenous immunoglobulin synthesis.

3.4.1 Immunization

For a typical fusion we immunize BALB/c female mice intraperitoneally with 10 to 100 μg of antigen in 100 μl of PBS and emulsified in 100 μl complete Freund's adjuvant. At intervals of at least 3 weeks, two intraperitoneal boosts are given with antigen emulsified in incomplete Freund's adjuvant. Between 3 and 4 days before the fusion, the mice are injected intravenously with 10 μg of antigen in 100 μl of physiological saline. Prolonging the time between boosts tends to result in higher affinity antibodies. We have obtained excellent mAbs from mice that had received the primary immunization over a year before the fusion.

We use a method of cell fusion that was loosely adapted from a technique described by Reading (12).

3.4.2 Materials

The standard medium for culturing the myelomas and hybridomas is Dulbecco's Modified Eagles Medium containing 4.5 g/litre glucose and supplemented with non-essential amino acids (Gibco), 1 mM sodium pyruvate, 100 units/ml penicillin, 100 μg/ml streptomycin, and 40 μM 2-mercaptoethanol (DMEM). Where indicated, the DMEM is further supple-mented with Fetal Calf Serum (FCS). It is best to either buy FCS pre-tested for its ability to support hybridoma growth or to test several different batches of FCS for their respective cloning efficiencies. Pre-tested polyethylene glycol

(1300–1600 Da; PEG) in 2 g sterile aliquots is available from the American Type Culture Collection (ATCC). Just before the fusion, the PEG is melted and diluted to 47% with prewarmed (37 °C) DMEM. The diluted PEG is kept in a 37 °C waterbath. HAT solution (100× concentrated) can be purchased from a number of suppliers including ATCC. On the day of the fusion, the HAT should be diluted with DMEM containing 30% FCS (DMEM/30% FCS/HAT).

3.4.3 Cell fusion

This is carried out as described in *Protocol 1*.

Protocol 1. Procedure for cell fusion in the preparation of monoclonal antibody

1. Sacrifice the mouse following criteria established by local animal care authorities. Place the dead animal briefly in a beaker of 70% ethanol while it is being transfered to a laminar flow hood to reduce the risk of contamination.

2. Aseptically remove the spleen and place it in a 100 mm Petri plate containing 10 ml of DMEM. Release the cells from the spleen by gentle teasing with two sterile pointed forceps. Pipette the cell suspension several times with a 10 ml pipette and filter the suspension using a nylon filter into a 50 ml sterile plastic culture tube. Wash the petri plate with 10 ml of DMEM and filter the wash into the culture tube containing the first 10 ml of cell suspension.

3. Centrifuge the spleen cells at 300g for 10 min. Resuspend the cells gently in about 200 µl of DMEM. After several minutes begin to slowly add additional DMEM with gentle agitation to a total volume of 20 ml. Centrifuge and resuspend the cells in 5 ml of DMEM, again initially suspending the cells in a small volume of medium (200 µl).

4. Suspend myeloma cells (we use SP2/O-Ag14), growing at a density of about 4×10^5 cells/ml in DMEM containing 5% (FCS) with greater than 95% viability, by pipetting and transfer to a 50 ml plastic tissue culture tube. Approximately $2–5 \times 10^7$ cells will be needed for the fusion.

5. Wash the cells twice with DMEM as described for the splenocytes and resuspend in 5 ml of DMEM.

6. Dilute the washed splenocytes and myeloma cells 1/100 and 1/20, respectively, in DMEM containing 5% trypan blue and count the viable cells in a haemocytometer.

7. Add the appropriate volume of the myeloma cell suspension to the spleen cells to give a ratio of spleen cells to SP2/0 of about 4:1. This ratio is not absolute. We have had successful fusions with ratios varying from 2:1 to 10:1.

Protocol 1. *Continued*

8. Centrifuge the cells for 10 min at 300g, remove the supernatant and loosen the pellet by flicking the tube without adding medium. Incubate the cells for about 1 min at 37 °C.

9. Add 1 ml of 47% PEG to the cells over 1 min with very gentle stirring using the pipette tip and then leave the cells for a further 1 min.

10. Add 2 ml of DMEM over 3 min with gentle stirring.

11. Add 7 ml of DMEM/30% FCS/HAT over 3 min and then slowly dilute the cells to 100 ml with DMEM/30% FCS/HAT.

12. Distribute the cells in 100 µl aliquots into ten 96-well microculture plates with the aid of an 8- or 12-channel pipette (Titertech, Flow Laboratories) and transfer the plates to a 37 °C incubator with an atmosphere of 5% CO_2/95% air.

13. On the next day, add 100 µl of DMEM/30% FCS/HAT to each well using the multi-channel pipette. Clones become apparent in the wells 3 days after the fusion.

3.4.4 Screening of hybridoma supernatants

When the clones occupy about 10% of the surface of the well (7–10 days after the fusion), the supernatant can be tested for the presence of specific antibody. We normally have growth in almost all of the wells. In certain fusions, the growth rate is similar in all wells of the fusion and all can be tested on the same day. This facilitates the screening, as the multi-channel pipette or other such devices can be used to take aliquots of supernatants. In other cases the growth rate is unequal and clones that are ready to be tested must be identified microscopically and aliquots of the supernatants removed individually.

We routinely use a solid phase antibody-capture assay to screen for antibody in culture supernatants. We and others have noted, however, that antigens can undergo conformational changes upon being adsorbed to polystyrene and, in some cases, one tends to selectively detect antibodies directed against altered conformational forms of the antigen. For certain fusions we have also used an antigen-capture assay in which immobilized anti-mouse immunoglobulin binds the mAb in the culture supernatant which can, in turn, bind radiolabelled antigen. Only the antibody-capture method will be described.

A number of positive and negative controls should be included in the screening. These would include an irrelevant mAb as a negative control and, if available, a mAb specific for the antigen as positive control. In addition, a number of randomly selected supernatants should be tested in Removawells (Dynatech) saturated with the BSA/PBS but not coated with antigen.

Putative positive clones should also be tested against wells not coated with antigen. In certain fusions we have found that all of the wells appeared initially to be positive. This was due to the presence of large numbers of antigen-specific plasma cells present in the cultures. Such plasmocytes can survive and secrete antibody in culture for over a week. In such cases we were forced to remove and replace the DMEM/FCS/HAT in all of the wells and, after 2 days, rescreen all of the supernatants.

Protocol 2. Screening for the presence of specific antibodies

1. Dissolve antigen in 5 mM glycine pH 9.2 (containing 0.02% NaN_3) and distribute in 50 µl aliquots into Immulon II Removawells (Dynatech). We use a concentration of 1–2 µg/ml for purified apolipoproteins and 30 µg/ml for LDL. Allow the antigen to adsorb to the plastic overnight at room temperature in a wet box (a sealed plastic box containing a water-saturated paper towel). The wells can then be used immediately or stored at 4 °C for up to two weeks.

2. Remove excess antigen and wash the wells four times with a solution of 0.015 M phosphate-buffered saline, 1 mM EDTA, 0.02% NaN_3 (PBS) containing 0.025% Tween 20 (PBS–Tween). This can be done using a wash bottle containing the PBS–Tween or by dipping the plates into a 4 litre beaker of PBS–Tween. After the last wash tap the wells dry on a paper towel.

3. Saturate the plastic by a 30 min incubation at room temperature with 1% bovine serum albumin in PBS (PBS–BSA). Then remove the PBS–BSA and tap the wells dry.

4. Remove aseptically 50 µl aliquots of the culture supernatants from the microculture wells and transfer to the antigen-coated Removawells. Return the Removawells to the wet box and incubate for a minimum of 2 h at room temperature.

5. Discard the culture supernatant and wash and tap dry the Removawells as above.

6. To detect specifically bound mAb, add 50 µl of diluted [^{125}I]anti-mouse IgG (Section 3.4.5) to each well. Incubate the wells for a minimum of 2 h at room temperature in a wet box, remove the mouse IgG and wash the wells as above. Appropriate care should be taken in handling and disposing of radioactivity.

7. Break apart the dried wells and place them in suitably-sized tubes for counting in a gamma spectrophotometer. We consider positive wells to have counts at least three times over background or in the range of the positive control.

3.4.5 Radioiodination of anti-mouse immunoglobulin

We use the modification (13) described in *Protocol 3* of the Chloramine T method (14) for labelling anti-mouse Ig. We normally achieve specific activities of $1-2 \times 10^4$ cpm/ng protein.

Protocol 3. Radiolabelling of anti-mouse immunoglobulin

1. The protein to be labelled: Dilute 50 µg of the protein to 170 µl with 0.3 M Na phosphate buffer (pH 7.0). For the antibody capture technique, we use affinity purified goat anti-mouse heavy and light chains (Jackson Immuno research). No serine protease inhibitors (e.g. phenylmethyl-sulphonyl fluoride) should be present.

2. Chloramine T: Prepare a solution of 90 µg/ml Chloramine T in 0.3 M sodium phosphate buffer (pH 7.0) immediately before the labelling.

3. ^{125}Iodine: Supplied as sodium iodide in dilute sodium hydroxide free from reducing agents, with a specific activity of about 14 mCi/µg and a concentration of 100 mCi/ml (Amersham).

4. Ion exchange column: The barrel of a 3 ml plastic syringe serves as the column. Pack loosely a small amount of silanized glass wool into the bottom of the syringe. Add a slurry of Dowex AG1-X8 (BioRad) in PBS to the plugged syringe to give a final packed volume of about 1.5 ml. It may be necessary to start the flow by using the syringe plunger. Wash the column with 20 ml of PBS followed by 5 ml of 0.1% BSA in PBS. The flow is stopped by placing the plunger in the syringe.

5. Reaction: Allow the reaction to take place on ice *inside a well-functioning fume hood with the front panel appropriately lowered and behind a lead glass shield.* Add 10 µl (1 mCi) of the stock solution of ^{125}I to the diluted protein followed by 20 µl (1.8 µg) of the chloramine T solution. After gentle mixing, allow the reaction to proceed on ice for 30 min and terminate by the addition of 20 µl of a 1 M KI solution.

6. Separation of free and protein-bound ^{125}I: Remove the plunger from the Dowex column and let the excess 0.1% BSA drain from the column. Immediately after the termination of the iodination reaction, add, with a Pasteur pipette, about 0.1 ml of 0.1% BSA to the reaction tube and, with the same pipette, transfer the contents to the top of the column. Elute with 0.1% BSA and collect the first 6 ml of eluant. The protein-bound ^{125}I is in the eluant whereas free ^{125}I is retained on the column.

3.4.6 Use of enzyme-linked immunosorbent assay (ELISA) for screening

The antibody capture method can be easily adapted to an ELISA format by substituting an appropriate enzyme–antibody complex for the [^{125}I]anti-mouse IgG. This most notably eliminates the necessity of working with radioactivity and is the method of choice of many investigators. The reader is referred to reference (1) for detailed protocols for screening using an ELISA protocol.

3.4.7 Subcloning

As irrelevant clones or cells that have lost the ability to secrete antibody may quickly overgrow the clones of interest, it is imperative to rapidly reclone the cells and identify those subclones that continue to secrete large amounts of specific antibody. Although we have occasionally subcloned in soft agar, we routinely use the limiting dilution method described below.

Protocol 4. Subcloning by the limiting dilution method

1. Cells from 96-well culture plates that were identified as being positive in the first screening are cultured sequentially in DMEM/30%FCS/HAT in 24-well plates, 30 cm petri plates, and finally in 100 mm petri plates. When sufficient cells are present at each step, test the culture supernatant for specific antibody as described above (*Protocol 2*)

2. When cells have reached a density of about 5 × 10^5/ml in the 100 mm petri plates they are resuspended with a pipette and transfered to a culture tube.

3. Count the cells in a haemocytometer using trypan blue exclusion to estimate viability as described above (*Protocol 1*).

4. Remove an appropriate aliquot and dilute to give 5 cells/ml in a total volume of 20 ml and, using a multi-channel pipette, redistribute 200 μl aliquots in each well of a 96-well culture plate (an average of 1 cell/well). Freeze the remaining cells in the culture tube (Section 3.4.8).

5. When sufficient growth has occurred, test the culture supernatants as above for the presence of specific antibody. Transfer the cells in the 12 wells giving the highest counts in the screening to individual wells of a 24-well plate. When there has been sufficient growth, 10^{-1}, 10^{-2}, and 10^{-3} dilutions of supernatant are tested for antibody activity and the cells producing the highest concentration of antibody are transfered to 100 mm petri plates.

6. Reclone the cells twice more using the same protocol but a density to give 0.5 cells/well. All subclones should be positive for specific antibody in the third recloning, a characteristic of monoclonality. The FCS supplement in the DMEM can then gradually be reduced from 30% to 5%.

3.4.8 Freezing and thawing of hybridomas

DMEM, FCS, and dimethylsulphoxide (DMSO) are mixed at a volume ratio of 80:10:10. The DMEM/10%FCS/10%DMSO is then sterilized by filtration using a 022 μm Millex-GV filter (Millipore). Cells at a density of about 5×10^5 cells/ml are resuspended with a pipette and transfered to a culture tube. Viability as determined by trypan blue exclusion should be greater than 90%. Cells are centrifuged for 5–10 min at 300g and slowly resuspended in a volume of DMEM/10%FCS/10%DMSO to give a density of about 2×10^6 cells/ml. Aliquots of 1.5 ml are transfered to sterile 1.8 ml cryotubes (Nunc) and left for 10 min at 37 °C. The tubes are transfered to a styrofoam box that contains a number of large pieces of metal (to retard cooling). The box containing the cells is then placed in −60 °C freezer.

Unlike the freezing of cells (which must be done slowly), cells must thawed quickly. The tube to be thawed is removed from the −60 °C freezer or from the liquid nitrogen and immediately transfered to a 37 °C water bath and continually shaken in the water to maximize heat transfer. When the cells are completely thawed, they are diluted to 10 ml with DMEM/10%FCS and centrifuged for 10 min at 300g. The supernatant is removed, the cells are slowly resuspended in DMEM/10%FCS, transfered to a culture vessel and are maintained as usual.

While we have kept certain clones at −60 °C for several years with good viability, this is not always the case. After several days it is best to transfer the frozen cells to liquid nitrogen. Frozen cells should be put back into culture and refrozen about once every two years to assure good viability. We maintain six frozen tubes of each clone, preferably distributed into two or more separate liquid nitrogen containers.

3.4.9 Production of ascites

When hybridomas are passaged as ascitic tumours in mice, concentrations of 2–10 mg/ml of mAb are present in the ascitic fluid and plasma. This is about 500 times those which we obtain in hybridoma culture supernatants.

We normally inject 5×10^6 hybridoma cells (in 0.5 ml DMEM without FCS) intraperitoneally into BALB/c female retired breeders. After 10 to 21 days, when abdominal swelling is evident, but before the mouse becomes overtly sick, the mouse is anaesthetized with ether, and exsanguinated by cardiac puncture (a 1-ml syringe with a 25 gauge $\times$ 5/8″ needle). The peritoneal cavity is opened and the ascitic fluid is harvested with a Pasteur pipette and pooled with the blood. Ascites formation can be accelerated by injecting the recipient mice with the tumour promoter, Pristane, several weeks prior to injection of the hybridoma (1). While many workers drain the ascites from the hybridoma-bearing mice several times by intraperitoneal puncture using a large bore needle, in our hands, this has not been efficient in terms of mAb yield. We obtain, on average, about 5 ml of pooled blood and ascites per mouse containing a total of about 40 mg of mAb. Pooling of blood with ascites increases the yield of antibody and the presence of the blood favours the formation of a compact clot. The serum/ascites mixture is prepared using the same protocol as that for serum preparation. Repeated freezing and thawing of ascites should be avoided.

3.5 Purification and characterization of mAbs

3.5.1 Purification of mAbs

mAbs should be characterized with respect to their immunoglobulin class and subclass and this can most easily be done using one of a number of commercially available kits (e.g. Biorad Laboratories, Calbiochem Corp. etc.). We use two methods for purifying IgG mAbs. The IgG subclass containing the mAb can be purified by Staphylococcal Protein A Sepharose (Pharmacia) by following the protocol of Ey et al. (15). If large quantities of IgG must be purified or, if many different mAbs must be simultaneously purified, we have found the caprylic acid/ammonium sulphate precipitation protocol of Reir et al. (16) to be advantageous and to give satisfactory purity for most uses.

3.5.2 Characterization of mAbs

Monoclonal antibodies against apolipoproteins should be characterized with respect to affinity, specificity, and 'pan-reactivity'. In addition, it is often of interest to know the intramolecular specificity of mAbs. While a detailed description of the various methods that can be used is beyond the scope of this chapter a number of approaches are listed below.

- Reactivity of mAbs with apolipoprotein isoforms (e.g. apo E isoforms) (17) or species (e.g. apo B48, apo B100) (18).
- Reactivity of mAbs with easily identifiable proteolytic fragments of apolipoproteins (e.g. thrombin or kalikrein fragments of apo B and CNBr fragments of apo AI) (18–20).
- Reactivity of mAbs with apolipoprotein fragments produced in bacterial expression systems (21,22).
- Reactivity of mAbs with synthetic peptides that mimic portions of apolipoprotein primary structure (17).
- Antibody competition (2) or cotitration (23) to determine if different mAbs against the same apolipoprotein react with the same or different epitopes.

4. Immunoreaction of apolipoproteins on replicas of gel electrophoresis

The ideal way to identify the apolipoproteins in a mixture is to concentrate and separate them by their electrophoretic properties, visualize all of them by staining, and identify individual apolipoproteins by specific immunoreaction on a replica of the electrophoresis that has been transferred to a nitrocellulose or polyvinylidene difluoride (PVDF) membrane. Depending on the sensitivity that is required, the staining can be performed in a duplicate electrophoresis gel, or on the replica membrane itself. The former is the more sensitive, especially with silver stain; 1 µg per band can be detected by staining with Coomassie Blue in the gel or Ponceau Red on an electrotransfer membrane, and about 1 ng by immunoreaction and autoradiography. We usually apply much more sample than this, in order to see all components (e.g. up to 50 µg apoHDL; more than that will overload the electrophoresis).

We usually make one of two kinds of comparisons: either (i) several samples are compared side by side in separate wells for the presence of one apolipoprotein antigen, or (ii) the immunoreaction of several antibodies or antisera are compared using one sample. For (ii) we use a well-former that has been cut to make one wide well and one narrow well to the side (for a standard or comparison sample), or sometimes no well at all is used—the sample is applied across the entire surface of the gel. After electrotransfer to a nitrocellulose membrane the replica can be cut into vertical strips of about 6–10 mm width (each one running from the cathode to the anode). Each of these strips can be tested with a different antibody, and since the sample was spread equally across all of them at the time of electrophoresis, electrophoretic mobility will be identical or uniform in each.

Slab gels in either standard (16 × 14 cm) or minigel (8 × 6 cm) size are suitable for electrophoresis before electrotransfer of replicas to membranes. Minigels have a clear advantage in rapidity of electrophoresis and less exposure of the sample to the gel components: this could be important if electrophoresis and electrotransfer are used to isolate proteins for subsequent immunization or amino acid sequence analysis in parallel with immuno-detection. Techniques for electrophoresis and isoelectric focusing were described in detail by Mills, Lane, and Weech (24).

4.1 Electrotransfer to membranes

We have found that the following technique captures about 80% of apolipoproteins, such as AI and D, from both SDS–PAGE and IEF gels on both pure nitrocellulose (e.g. Schleicher and Schuell BA85) and polyvinyl-idenedifluoride (Millipore Immobilon PVDF) membranes. Nitrocellulose membrane absorbs water more easily than PVDF and is the better choice for immunoreactions. PVDF is more resistant to tearing, but its prime advantage is inertness to the reagents used in automated Edman degradation for amino acid sequencing, allowing one electrophoresis replica to be used for both immunochemical and primary structure characterization of apolipoproteins.

To prepare electrotransfer buffer 60 mM boric acid is adjusted to pH 8.0 with NaOH (this can be stored as a 10× concentrated solution). For use, mix seven volumes of the sodium borate with three volumes of methanol. Prepare the gel-membrane sandwich in a large shallow dish, with 1–1.5 cm of water at the bottom, in the following order: plastic cassette side, wet sponge, wet blotting paper (equivalent in thickness to Whatman 3MM). Carefully lower the dry nitrocellulose onto the wet blotting paper, avoiding trapping bubbles. Alternatively, wet the PVDF with methanol and then with water and immediately place it on the blotting paper. Immediately lower the electrophoresis gel onto the wet membrane, eliminating any bubbles. Place a wet blotting paper and sponge on the top and then cover with the other side of the plastic cassette.

Three-quarters fill the electrotransfer chamber with borate–methanol solution and put the gel-membrane sandwich in, ideally only one sandwich per chamber to minimize contamination and so keep the background as clear as possible. Electrophorese the proteins from the gel to the membrane towards the anode (positive). Best results are obtained at 100 mA per chamber for 12–16 h. Alternatively a current of 400 mA for 3 h may be used. About 1 µg protein per band can be detected on the membrane using 0.2% Ponceau Red in 3.5% trichloracetic acid for 5 min, with destaining in water and 10 mM Tris, 0.15M NaCl, 0.02% NaN_3 pH 7.4.

4.2 Immunoreaction of electrophoresis replicas

Prepare 20 litres of 10 mM Tris, 0.15 M NaCl, 0.02% NaN_3, pH 7. Saturate protein-binding sites on the membrane with 3% (w/v) polyvinylpyrrolidone, molecular weight 44 000 (PVP) in the Tris–NaCl buffer, for 1 h. Incubations can be made at room temperature or preferably at 37 °C in a sealed polyethylene bag. Incubate the membrane with antibody diluted in 1% bovine serum albumin (BSA) in Tris–NaCl buffer for 90 min. The optimum dilution is the most dilute that gives maximum reaction of the specific bands, and minimum background. This is usually $1:10^3–1:10^6$ depending on the individual antiserum or ascites. Rinse the membrane six times in Tris–NaCl buffer. Incubate the membrane with 10^6 c.p.m. ^{125}I-labelled IgG against mouse IgG, for 1 h. Rinse the membrane six times in Tris–NaCl, let it dry and arrange the membranes on a sheet of blotting paper, fixing them in place with adhesive tape. Place this sheet in a polyethylene bag and (in a dark room) expose it to Kodak XAR-5 film in a metal holder, with the film between the bag and a Cronex Lightning Plus Intensifying screen (DuPont). exposure is usually sufficient. The advantages of autoradiography are the development is usually sufficient. The advantages of autoradiography are the intense black bands and high contrast that can be obtained, and the ease of making shorter or longer exposures of the same membrane. If backgrounds are excessively dark, the results can often be improved by using the IgG of the first antibody (isolated by chromatography on Protein A Sepharose) in place of ascites or antiserum, or by radioiodinating this antibody itself.

If many analyses are to be performed using the same antibodies, it is worthwhile to use a modification of the incubation conditions given above to make a single step reaction. The saturation and rinsing of the membrane are carried out as described above, but the incubation is performed in a mixture of the diluted first antibody and ^{125}I-labelled IgG against mouse IgG-Fc. The optimum dilution of first antibody is found by incubating a series of identical strips of electrophoresis replica with mixtures of 10^6 c.p.m. labelled anti-IgG-Fc and one of a series of 12 two-fold dilutions of first antibody, starting from 1:1000, each in a constant volume. The dilution that gives the maximum intensity specific band has the optimum ratio of first antibody to labelled anti-IgG-Fc and should be used for subsequent experiments.

5. Immunoassays of apolipoproteins

A number of different immunoassays are being used to measure apolipo-proteins. These include single radial immunodiffusion, electroimmuno-assay, immunonephelometry, radioimmunoassay, radioimmunometric assay, and ELISA. Some assays are commercially available in kit form. At present, however, there is considerable inter-method and inter-laboratory variation in apolipoprotein determinations. This variability is in part a function of the standards used in the assay and in part inherent in the methodologies themselves. Several national and international bodies are attempting to establish guidelines for the standardization of apolipoprotein determinations. Reports from the International Federation of Clinical Chemistry (25) and from the Apoprotein and Antibody Standardization Program Planning Committee of the National Heart, Lung, and Blood Institute (26) have been recently published. These include recommendations concerning the collections handling, and storage of samples, the selection and characterization of antibodies, methodologies, and the choice and handling of primary and secondary standards and references.

We will present the protocol used in our laboratory for the measurement of apo AI by radioimmunometric assay using an anti-apo AI mAb (7).

5.1 Radioimmunometric assay

5.1.1 Materials

- Coating buffer: 50 mM Sodium carbonate, pH 9.6
- Wash buffer: PBS (8 g NaCl, 0.2 g KCl, 1.44 g $Na_2 HPO_4$, 0.24 g KH_2PO_4/ litre, pH 7.2) containing 0.05% Tween 20
- Saturation buffer: 0.5% gelatin in PBS
- Dilution buffer: 0.5% gelatin in PBS containing 0.05% Tween 20
- Standard Serum: Serum whose apo AI content has been determined by the isotope dilution technique (Section 5.2)

Protocol 5. Procedure for antibody titration and determination of coating concentration

1. Arrange Immulon II Removawells in holders. Coat two rows of 12 wells with 5 µg/ml apo HDL (total protein), two rows with 2 µg/ml, two rows with 1 µg/ml, and two rows with coating buffer. Leave at room temperature in a wet box for 2 h.

2. Discard the solutions, wash wells once with PBS, blot dry, and saturate the wells by adding 250 µl of saturation buffer to each well. Leave in a wet box for 30 to 60 min at room temperature.

3. During the coating and saturating steps, prepare 11 doubling dilutions of the anti-apo AI mAb in dilution buffer starting at a dilution of 1/500.

4. Discard the saturating solution, wash wells once with PBS, blot dry, and add 100 µl of the antibody dilutions. To the last wells in each row, add 100 µl of dilution buffer. Incubate for 1 h at room temperature.

5. Discard the antibody solution, wash three times with wash buffer and blot dry.

6. Add 100 µl of [^{125}I]anti-mouse Ig, diluted to 2×10^5 c.p.m./100 µl with dilution buffer to all wells. Incubate in a wet box for 1 h at room temperature.

7. Discard the radioactivity and wash wells three times with wash buffer. Blot dry and determine bound radioactivity.

8. The concentration of apo HDL to be used in the assay is the highest dilution that gives maximal counts. The concentration of mAb to be used is two times that which gives 60% of maximum binding because, in the actual competitive assay, the mAb is diluted 1:1 with the sample.

Protocol 6. Procedure for competitive radioimmunometric assay

1. Prepare sufficient apo HDL-coated and gelatin-saturated-Immulon II Removawells using the appropriate dilution (*Protocol 5*, step 8) of apo HDL in coating buffer by following steps 1 and 2. Wash once with PBS and tap dry.

2. In 96-well microtitre plates prepare:
   (a) seven doubling dilutions (100 µl) of standard plasma in duplicate starting with a dilution of 1/160,
   (b) three doubling dilutions (100 µl) in duplicate of each plasma sample to be tested starting at 1/40,
   (c) three doubling dilutions (100 µl) of HDL samples to be tested starting at 1/80

Protocol 6. *Continued*

    (d) six wells containing 100 μl dilution buffer (for maximum binding), and

    (e) four wells with 200 μl dilution buffer (second antibody negative control to estimate background).

3. To each column of wells in the microtitre plate, add 100 μl of appropriately diluted mAb (*Protocol 5*, step 8) with an 8-channel pipette, mix by pipetting four times and transfer 100 μl immediately to the apo HDL-coated Removawells. Repeat with the next column and until all the wells are transfered. Place the Removawell plates in a wet box and incubate with gentle shaking for 1 h at room temperature.

4. Continue as described in *Protocol 5* with steps 4 through 7. It should be noted that the time needed for manual delivery of reagents in each plate from row 1 through 12 could influence results in a short assay as described here if strict chronometry of sample addition were not followed. We follow a strict timing for each step such as loading one plate every 5 min. This will limit a skilled operator to an assay that includes eight microtitre plates.

5. The results are expressed as (c.p.m. − background)/(maximum c.p.m. − background) or B/Bo. B/Bo is plotted against the log apo AI concentration for the standard serum. The points in the middle linear portion of the curve are used for linear regression. The B/Bo values of the samples are used to extrapolate the concentration of apo AI in the diluted samples. These values are then adjusted for the sample dilution.

5.2 Isotope dilution

It seems that, in general, pure isolated apolipoproteins do not express antigenic sites in the same way as lipoproteins, and further, that lipoproteins are heterogeneous in their expression of antigenic sites. The preferred standard for comparison of immunoassays between laboratories is currently a common plasma sample. This has usually been used only as a relative or secondary standard, but we have developed a method by which plasma can be used as an independant primary standard. We have applied the method of isotope dilution to measure apo AI and apo AII (27, 28) but in essence it could be applied to any protein that can be isolated, and certainly other techniques of isolation such as two-dimensional PAGE and electrotransfer can be effectively employed. The method allows us to measure the amount of apo AI in a few ml of a plasma sample; many aliquots of about 200 μl of the remaining plasma can be stored at −70 °C to be used as standards in immunoassay.

    The principle of the method is that the molecule of interest is isolated from the unknown sample and the amount of pure product is measured. In order to correct for losses of the compound throughout the isolation a known amount of a tracer is added at the start, the fraction of the tracer that is recovered is calculated and the value is used to estimate the amount of the unknown compound that would have been obtained if the yield was 100%. If a significant amount of tracer was used and it contributes to the measurement of the unknown, then the final correction is to subtract its contribution. We will describe our technique for apo AI (27). ^{125}I-labelled apo AI is used as the tracer and an extra isolation step from plasma is added following the labelling. The purpose of this step is to remove any denatured apo AI that may have formed during the labelling, since Osborne *et al.* (29) observed that some iodo-apo AI self-associated irreversibly.

5.2.1 [125I]apo AI

Isolate HDL from plasma by ultracentrifugation, dialyse it against 10 mM NH_4HCO_3, 1 mM Na_2 EDTA, 0.02% NaN_3 to remove salt, lyophilize, and delipidate it with chloroform–methanol 2:1 (24, 30) and then resolubilize the protein with 0.1% SDS. Chromatograph about 5 mg of the apo-HDL on a gel permeation column of Sephacryl S300 (Pharmacia) (1.5 × 100 cm) in 0.1% SDS, 0.1 M NaH_2PO_4, 0.02% NaN_3, pH 7.0. Monitor the elution of protein at 280 nm and by SDS–PAGE and pool the fractions containing pure apo AI. These are usually the central fractions in the apo AI peak. Dialyse the apo AI against 0.15 M NaCl, 0.05 M NaH_2PO_4, 1 mM EDTA, pH 7.5 and label 1 mg/ml with about 0.5 mCi ^{125}I using chloramine T (24). Remove the free iodide by gel filtration through Sephadex G-25 (Pharmacia) in PBS (1 × 20 cm).

5.2.2 Screening of the [125I]apo AI

Add about 10^7 c.p.m. [125I]apo AI to each of several 3 ml aliquots of plasma, mix gently, and reisolate the HDL by density gradient ultracentrifugation, using 3 ml per tube, without staining (27, 31). This HDL, containing labelled apo AI that has exchanged with unlabelled apo AI, is the tracer for the isotope dilution measurements and its specific activity (c.p.m./mg apo AI) must be measured. Take the HDL that is equivalent to 3 ml plasma, dialyse it against the NH_4HCO_3 solution above, lyophilize, and redissolve it in 400 μl SDS gel filtration buffer above, but containing 1% SDS. Warm the sample to 60 °C for 5 min. Filter the sample through a small, low-protein-binding, 0.22 μm filter, and chromatograph 50–100 μl of it through a TSK 3000 (Pharmacia) HPLC gel filtration column (7.5 × 600 mm). Pool the fractions containing pure apo AI (usually the peak fractions from the main peak), take aliquots from this pool and measure the radioactivity and protein in each. This gives the specific activity SAo (c.p.m./mg apo AI) of the tracer.

5.2.3 Isotope dilution measurement

Add $2.5-5 \times 10^5$ c.p.m. of the radioactive tracer HDL to 3 ml of the test plasma, mix gently, and remove several aliquots of 20 µl for accurate determination of the total radioactivity added to the entire plasma sample—this is Ao c.p.m., and the mass of labelled apo AI added can be calculated from Ao and the specific activity SAo.

Isolate the HDL from this plasma by density gradient ultracentrifugation, then dialyze, lyophilize, and redissolve the HDL, and isolate pure apo AI by HPLC gel filtration as described in Section 5.2.2. Measure the specific activity SAr (c.p.m./mg) of this apo AI. This value will be less than that of the screened HDL due to the dilution by the unlabelled apo AI in the unknown sample.

The amount of apo AI in the unknown sample can be calculated from the relationship:

$$Mu = (Ao/SAr) - Mo$$

where Mu is the amount of apo AI in the unknown sample, Ao is the radioactivity added to it, Mo is the mass of the tracer radioactive apo AI added, and SAr is the specific activity of the recovered apo AI.

5.3 Other radioimmunometric assays

We use several variations of this format for measuring other plasma apolipoproteins. For the apo B assay, LDL (30 µg/ml) was used as the immobilized antigen and the antibody-antigen mixture was allowed to react in the wells for 24 h in the wells before addition of [^{125}I]anti-mouse Ig (32). During development of the cholesteryl ester transfer protein (CETP) assay, we discovered that a plasma component, probably apo AI, could bind to the solid phase CETP giving a competition curve resembling that produced by true CETP. This interference could only be eliminated by high concentrations of detergent such as 0.5% Triton (33). The substitution of purified apo AII for apo HDL as the solid phase antigen in the apo AII assay reduced the variation and increased the sensitivity of the assay. A similar immunoassay for apo D has also been developed where the semi-purified antigen is coated at 2 µg/ml and where the antigen–antibody reaction lasts 5 h in the absence of detergent (34). In summary, the optimal conditions for different antibodies and for each different assay must be established.

6. Immunoprecipitation of lipoproteins

Immunoprecipitation of radiolabelled lipoproteins is useful for studying the heterogeneity, the biosynthesis, and the assembly of lipoproteins and for evaluating antibodies as potential reagents for immunoassays. Normally, radioiodinated lipoproteins are used for studying lipoprotein heterogeneity and for evaluating the 'panreactivity' of antibodies whereas biosynthetically labelled proteins are used for studies of lipoprotein synthesis and assembly.

We use indirect immunoprecipitation by killed and fixed *Staphylococcus aureus* Cowan I strain (Calbiochem Corp.) The protein A of the *S. aureus* binds with low affinity or not at all to certain immunoglobulin classes and subclasses (15). Therefore, in the case of mouse mAbs of the IgM, IgA, or IgG1 isotypes, it is necessary to prearm the Pansorbin with rabbit anti-mouse immunoglobulin (2).

6.1 Preparation of armed Pansorbin

Packed *S. aureus* cells (2 ml) (Pansorbin is supplied as a 10% suspension) are washed three times (10 min, 10 000g) with 1% BSA in PBS. The washed Pansorbin is incubated for 24 h at 4 °C with 1 ml of rabbit anti-mouse immunoglobulin serum (Cederlane Laboratories) diluted 1/20 with PBS/BSA. The cells are then washed three times with 1% PBS/BSA and resuspended in a total volume of 20 ml PBS/BSA. The armed Pansorbin can be kept at 4 °C for at least 6 months but should be washed as above immediately prior to usage.

6.2 Immunoprecipitation

6.2.1 Immunoprecipitation of radioiodinated lipoproteins

[125I]Lipoprotein (25 ng protein) (35) with a specific activity of about 500 c.p.m./ng is mixed with a series of dilutions of mAb with both mAb and lipoprotein being diluted in PBS/BSA. The total reaction mixture is 0.55 ml. The mixture is incubated for 24 h at 4 °C at which time a 0.1 ml aliquot of a washed 10% armed Pansorbin suspension is added. The mixture is incubated for a further 24 h, centrifuged at 10 000g for 10 min. The pellet is washed once with PBS/BSA, centrifuged, and pellet-associated radioactivity determined. In parallel, an equal amount of the [125I]lipoprotein is precipitated with 12% trichloroacetic acid (TCA). Results are expressed as a percentage of TCA precipitable radioactivity. Incubation of the [125I]lipoprotein with an irrelevant mAb should be included to estimate non-specific precipitation by the armed Pansorbin.

6.2.2 Immunoprecipitation of biosynthetically labelled lipoproteins

Essentially, the same protocol can be used to immunoprecipitate biosynthetically labelled lipoproteins or apolipoproteins from cell lysates or from culture supernatants. To minimize non-specific precipitation, the sample should be precleared by a preincubation (30 min at 4 °C) with armed Pansorbin, followed by centrifugation. The supernatant is transfered to another tube and incubated overnight at 4 °C with 0.1 µl mAb (ascites). The complexes are then precipitated with armed Pansorbin and the pellet is washed. If the sample is to be analysed by SDS–polyacrylamide gel electrophoresis, the pellet is incubated for 10 min at 85 °C in 50 µl of 2% SDS, 10% glycerol, 100 mM dithiothreitol, 60 mM Tris pH 6.8, and 0.0001% bromophenol blue.

7. Immunoaffinity chromatography

Lipoprotein particles are heterogeneous with respect to size, density, and lipid and apolipoprotein composition. Immunoaffinity chromatography has been used to separate proteins on the basis of their apolipoprotein composition and epitope expression. As a consequence, one has been able to identify distinct subpopulations of particles within the same lipoprotein density subfraction that may differ with respect to structure, origin, metabolism, function, and clinical relevence.

Recovery and particle integrity are two problems associated with the immunoaffinity chromatography of lipoproteins. We have found that particles of the IDL and LDL size are close to the exclusion size of Sepharose 4B and tend to get trapped within the matrix of the beads. This problem is not appreciably improved by using Sepharose 2B as the matrix. As a result, recovery of particles in this size range is not good. This seems to be less of a problem with larger VLDL and smaller HDL. In certain cases we have improved recoveries by reversing the flow of the column for the elution of the retained fraction.

Not surprisingly, problems associated with maintaining the integrity of the lipoproteins are mainly with the fraction that is retained on the column. In this respect, we have found that elution using an acid pH is superior to that using chaotropic agents, although recovery may be somewhat less. Our experience with affinity chromatography of lipoproteins has largely been for the separation of apo B-containing lipoproteins of intestinal and hepatic origins, respectively (32). The following protocol is based on this experience.

While we normally activate Sepharose with CNBr in the laboratory (36), it is often more convenient to use commercially available preactivated Sepharose (e.g. Pharmacia) for preparing small quantities of immunosorbant.

Protocol 7. Procedure for the preparation of immunoadsorbants

1. Preparation of antibodies: For most purposes, a partial purification of the antibody by ammonium sulphate precipitation is sufficient.
   (a) Dilute ascites 1/4 v/v with PBS.
   (b) Add an equal volume of saturated ammonium sulphate (4.1 M, pH 6.8) slowly with stirring to the diluted ascites (final solution is 50% saturated with respect to ammonium sulphate). Allow precipitation to proceed for 4 h at 4 °C.

(c) Centrifuge for 15 min at 3000g, discard the supernatant and redissolve the precipitate in a volume of PBS equivalent to 50% of the starting undiluted ascites.

(d) Dialyse against 0.1 M NaHCO$_3$, pH 8.0.

2. Preparation of Sepharose: 1 ml of packed Sepharose will be required for each 4 mg of protein to be coupled.

(a) Wash the Sepharose on a Buchner funnel with 10 times the volume of ice-cold distilled water.

(b) Remove the water, leaving the gel moist.

(c) Transfer the gel to a beaker containing a magnetic stirring bar and add an equal volume of distilled water and two times the volume of 2 M Na$_2$CO$_3$. Place the beaker into an icebath positioned on a magnetic stirrer in a well functioning fume hood.

3. Activation of Sepharose:

(a) In the fume hood, prepare a 2 g/ml solution of CNBr in acetonitrile, the total volume being 1/10 that of the Sepharose to be activated.

(b) Add the CNBr solution to the gel. The CNBr will immediately precipitate. Continue stirring until the CNBr redissolves.

(c) Transfer the slurry to a Buchner funnel and wash with 20 times the volume of ice-cold H$_2$O and 20 times the volume of ice-cold 0.1 M NaHCO$_3$.

4. Coupling the protein:

(a) Drain and transfer the moist gel to a beaker containing a stirring bar and add the dialysed protein solution and an equal volume of NaHCO$_3$.

(b) Stir overnight at 4 °C and, next day, filter the gel on Buchner filter and collect the filtrate.

(c) Transfer the gel to a beaker, add two times the volume of a 1 M glycine solution, pH 8.0, and incubate with stirring at 4 °C for a further 4 h.

(d) Wash the gel by repeated cycles of 0.1 M Na$_2$CO$_3$, pH 10 and acetic acid, pH 3.5.

(e) Re-equilibrate the gel in PBS. The efficiency of coupling can be estimated by comparing the protein recovered in the filtrate with the starting protein. The gel can be used for batch immunoadsorptions or in the form of an immunoaffinity column.

7.1 Affinity chromatography

Before beginning experiments the column should be verified to assure that no antibody is leaching from the column either with neutral buffer or with the elution buffer. This can be done by testing samples of column washes using the antibody capture assay (Section 3.4.4). If antibody is present, the column should be washed until antibody can no longer be detected. The capacity of the column should be estimated by applying an excess of antigen and determining the difference between the quantity applied and that which was not retained.

Branch the column to a peristaltic pump set at a speed of 5 ml/h. When the column is ready, remove the top of the column, allow the buffer to enter the gel and manually apply the sample with a Pasteur pipette. After the sample has drained into the gel the column wall is washed with a small quantity of buffer. This is again allowed to drain into the gel and the column is filled with buffer attached to a buffer reservoir. A flow rate of 5 ml/h is maintained and fractions having a volume equivalent to the applied sample volume are collected (if one is interested in the non-retained fraction). The fractions should be monitored at 280 nm. When the optical density has returned to baseline, the column is washed with one column volume of 1 M NaCl (50 ml/h) to detach non-specifically bound lipoprotein. The bound proteins are then eluted with 0.1 M citric acid/1 M NaCl, pH 3.0, with a flow rate of 50 ml/h. The tubes containing the fractions with the highest optical density in both the retained and non-retained fractions are respectively pooled and dialysed against PBS. When several column volumes of the low pH elution buffer have been passed, the column is reequilibrated with PBS.

Because of the heterogeneity of the lipoproteins with respect to epitope expression, it is possible that an apolipoprotein will not be quantitatively retained by the immunosorbant even if the capacity of the column is not exceeded. This can be a problem, especially with immunosorbants made with mAbs. This can, in some cases, be overcome by preparing a column with several different mAbs. As an example, two different anti-apo B100 mAbs were required to completely separate apo B100 VLDL from apo B48 VLDL (32).

Acknowledgements

We wish to thank Louise Blanchette, Rino Camato, Xavier Collet, Helena Czarnecka, Philippe Douste-Blazy Maire de Lourdes, Mireille Hogue, Debbie Jewer, Lorraine Leblond, Roger Maurice, Peter Milthorp, Thanh Dung N'Guyen, Carole Poudrier, Elemer Raffai, Francois Tercé, Richard Théolis, Nathalie Tremblay, Roy Verdery, Camilla Vézina, and Zbigniew Zawadzki who were instrumental in developing in our laboratory the methods that are described in this chapter.

References

1. Harlow, E. and Lane, D. (1988). *Antibodies: A Laboratory Manual*, Cold Spring Harbor Laboratory Press, Cold Spring Harbor, NY.

2. Milne, R. W., Blanchette, L., Théolis Jr, R., Weech, P. K., and Marcel, Y. (1987). *Mol. Immunol.*, **24**, 435.
3. Marcel, Y. L., Hogue, M., Weech, P. K., and Milne, R. W. (1984). *J. Biol. Chem.*, **259**, 6952.
4. Schonfeld, G., Patsch, W., Pfleger, B., Witztum, J. L., and Weidman, S. Q. (1979). *J. Clin. Invest.*, **64**, 1288.
5. Milthorp, P., Weech, P. K., Milne, R. W., and Marcel, Y. L. (1986). *Arteriosclerosis*, **6**, 285.
6. Curtiss, L. K. and Smith, R. S. (1987). In *Proceedings of the Workshop of Lipoprotein Heterogeneity*, NIH Publication, 87–2646, p. 363.
7. Marcel, Y., Jewer, D., Leblond, L., Weech, P. K., and Milne, R. W. (1989). *J. Biol. Chem.*, **264**, 19942.
8. Ellouz, F., Adam, A., Ciorbaru, R., and Lederer, E. (1974). *Biochem. Biophys. Res. Commun.*, **59**, 1317.
9. Muller, R. (1980). *J. Immunol. Meth.*, **34**, 345.
10. Fievet, C., Durieux, C., Milne, R., Delaunay, T., Agnani, G., Bazin, H., Marcel, Y., and Fruchart, J. C. (1989). *J. Lipid Res.*, **30**, 1015.
11. Littlefield, J. W. (1964). *Science*, **145**, 709.
12. Reading, C. L. (1982). *J. Immunol. Meth.*, **53**, 261.
13. Mellman, I. S. and Unkless, J. C. (1980). *J. Exp. Med.*, **152**, 1048.
14. Hunter, W. M. and Greenwood, F. C. (1962). *Nature*, **194**, 495.
15. Ey, P. L., Prowse, S. J., and Jenkin, C. R. (1978). *Immunochemistry*, 15, 429.
16. Reir, L. M., Maines, S. L., Ryan, D. E., Levin, W., Bandiera, S., and Thomas, P. (1987). *J. Immunol. Meth.*, **100**, 123.
17. Weisgraber, K. H., Innerarity, T. L., Harder, K. J., Mahley, R. W., Milne, R. W., Marcel, Y. L., and Sparrow, J. T. (1983). *J. Biol. Chem.*, **258**, 12 348.
18. Marcel, Y. L., Hogue, M., Thèolis Jr, R., and Milne, R. W. (1982). *J. Biol. Chem.*, **257**, 13165.
19. Marcel, Y. L., Innerarity, T. L., Spilman, C., Mahley, R. W., Protter, A. A., and Milne, R. W. (1987). *Arteriosclerosis*, **7**, 166.
20. Weech, P. K., Milne, R. W., Milthorp, P., and Marcel, Y. L. (1985). *Biochim. Biophys. Acta*, **835**, 390.
21. Krul, E. S., Kleinman, Y., Kinoshita, M., Pfleger, B., Oida, K., Scott, J., Law, A., Pease, R., and Schonfeld, G. (1988). *J. Lipid Res.*, **29**, 937.
22. Pease, R. J., Milne, R. W., Jessup, W. K., Law, A., Provost, P., Fruchart, J.-C., Dean, R. T., Marcel, Y. L., and Scott, J. (1990). *J. Biol. Chem.*, **265**, 553.
23. Fisher, A. G. and Brown, G. (1980). *J. Immunol. Methods*, **39**, 385.
24. Mills, G. L., Lane, P. A., and Weech, P. K. (1984). *Laboratory Techniques in Biochemistry and Molecular Biology*, (ed. R. H. Burdon and P. H. van Knippenberg). Vol. 14, p. 171. Elsevier, Amsterdam.
25. Albers, J. J. and Marcovina, S. M. (1989). *Clinical Chem.*, **35**, 1357.
26. Albers, J. J. (1989). *Arteriosclerosis*, **9**, 144.
27. Weech, P. K., Jewer, D., and Marcel, Y. L. (1988). *J. Lipid Res.*, **29**, 85.
28. Weech, P. K., McConathy, W. J., Alaupovic, P., and Fesmire, J. (1980). In *Atherclerosis V. Proceedings of the Fifth International Symposium on Arteriosclerosis*, (ed. A. M. Gotto, Jr., L. C. Smith, and B. Allen), p. 804. Springer Verlag, New York.

83

29. Osborne, J. C., Schaefer, E. J., Powell, G. M., Lee, N. S., and Zech, L. A. (1984). *J. Biol. Chem.*, **259**, 347.
30. Olofsson, S. O., McConathy, W. J., and Alaupovic, P. (1978). *Biochemistry*, **17**, 1032.
31. Terpstra, A. H. M., Woodward, C. J. H., and Sanchez-Muniz, F. J. (1981). *Anal. Biochem.*, **111**, 149.
32. Milne, R. W., Weech, P. K., Blanchette, L., Davignon, J., Alaupovic, P., and Marcel, Y. L. (1984). *J. Clin. Invest.*, **73**, 816.
33. Marcel, Y. L., McPherson, R., Hogue, M., Czarnecka, H., Zawadzki, Z., Weech, P. K., Whitlock, M. W., Tall, A. R., and Milne, R. W. (1990). *J. Clin. Invest.*, **85**, 10.
34. Camato, R., Marcel, Y. L., Milne, R. W., Lussier-Cacan, S., and Weech, P. K. (1989). *J. Lipid Res.*, **30**, 865.
35. Bilheimer, D. W., Eisenberg, S., and Levy, R. I. (1972). *Biochim. Biophys. Acta*, **260**, 212.
36. March, S. C., Parikh, I., and Cuatrecasas P. (1974). *Anal. Biochem.*, **60**, 149.

The separation and analysis of high-density lipoprotein (HDL) and low-density lipoprotein (LDL) subfractions

E. ROY SKINNER

1. High-density lipoproteins: Introduction

Plasma HDL performs a wide variety of functions. In addition to its central role in lipoprotein metabolism (1) in which it provides a reservoir for small apoproteins and is implicated in the removal of surface material during the lipolysis of triglyceride-rich lipoproteins, HDL is intimately associated with the process of reverse cholesterol transport (2). In this pathway, it provides an acceptor for cholesterol effluxed from peripheral cells and serves as a carrier of this material to other lipoproteins for hepatic uptake and excretion. HDL provides steroidogenic and other tissues with a source of cholesterol by means of specific receptors (3). It is also reported to promote endothelial cell proliferation (4), to protect LDL against oxidation (5), and reduce the rate of post-prandial lipaemia (6). With such a diversity of functions, it is not surprising that HDL is highly speciated. The similarity in the composition of the various subspecies, however, has made their separation a task that has, until recently, largely eluded the attempts of most investigators; the current belief that the interactions of subspecies is likely to account for the protective effect of HDL against atherosclerosis (7) has provided a considerable stimulus for further analytical studies.

HDL is normally considered to consist of those lipoprotein particles which fall within the density range 1.063 to 1.21 g/ml. Within this density region are found a variety of subspecies of HDL, each of characteristic particle size and apoprotein and lipid composition. Because the addition or removal of constituents occurs in a manner which is consistent with the maintenance of a thermodynamically stable particle, each step in the metabolic interconversion of subspecies occurs as a 'quantum jump', resulting in the presence in the plasma of a series of discrete particles rather than a continuous spectrum.

HDL contains two major subfractions, HDL_2 and HDL_3, of densities 1.063 to 1.125 g/ml and 1.125 to 1.21 g/ml, respectively. Lipoprotein particles with the characteristics of HDL but with densities in the region 1.055 to 1.063 g/ml are also present in the plasma of some subjects (referred to as HDL_1 (8)) and in certain animal species that have been subjected to a diet high in cholesterol (designated HDL_c (9)). HDL_2 and HDL_3 have each been demonstrated to contain more than ten fractions by the use of a combination of isoelectric focusing and immunological techniques (10) that are not appropriate for routine assay. The resolution of HDL into five subpopulations by gradient gel electrophoresis (11) has provided a useful system for the definition of some of these components. This section will describe methods that are available for the isolation and analysis of HDL subfractions.

2. The separation of total HDL from plasma

2.1 Preparative ultracentrifugation

The separation of HDL into subspecies usually necessitates the isolation of the total HDL fraction from plasma as a first step and this is normally achieved by flotation in the preparative ultracentrifuge. A number of special considerations arise in the analysis of HDL in addition to those described in Chapter 1 for the general separation of lipoproteins and the procedure will therefore be given in detail. It is strongly recommended to use fresh and non-frozen samples of plasma to avoid the occurrence of changes in the composition and properties of the HDL subfractions; in some cases, however, (e.g. when subsequent analysis is to be made by gradient gel electrophoresis only), satisfactory results can be obtained with plasma that has been stored at $-20\,°C$ for up to two months. Since the strong salt solutions and high gravitational forces that are inherent in the method result in substantial dissociation of some apolipoproteins (apo AIV, apo E, and apo C) from the native HDL particles (12, 13, 14), it is vital that standard conditions, especially centrifugation times, are strictly adhered to in order that a uniform product may be obtained from different preparations. Allowance must be made for this loss in interpreting data, especially in metabolic studies, since apolipoproteins play a crucial role in the interconversion and tissue uptake of HDL particles. The method may be applied quantitatively with recoveries of approximately 92–96%.

2.1.1 Reagents

A series of NaBr solutions of different densities are prepared by dissolving the amounts of NaBr and EDTA shown in *Table 1* in distilled water and making up the volume to 1 litre with distilled water. Measure the density of the solution at 20 °C using a density meter (e.g. Anton Paar DMA Digital Density Meter) and adjust the density to the required value by adding solid

Table 1. Preparation of NaBr solutions of different densities

Solution	Required density (d_1)	Weight of NaBr (g)	Weight of Na_2EDTA (g)	Final volume (ml)
(a)	1.063	7.6	0.37	1000
(b)	1.117	13.8	0.37	1000
(c)	1.187	20.9	0.37	1000
(d)	1.210	23.2	0.37	1000
(e)	1.295	30.5	0.37	1000
(f)	1.357	34.2	0.37	1000

NaBr or distilled water. The most convenient way to prepare a solution of a given density is to make up the solution rather more dense than that required and to calculate the volume of water that must be added to reduce the density to the required value using the formula:

$$x = \frac{d_2 - d_1}{d_1 - 1}$$

where x = volume (ml) of distilled water to be added to a 1 ml volume of the NaBr solution to give the required density, d_1 = required density, and d_2 = measured density.

Protocol 1. The separation of total HDL from plasma by flotation in the preparative ultracentrifuge

1. Collect blood samples and mix in a dry tube containing Na_2EDTA (final concentration 1 mg/ml). Alternatively, draw the blood directly into Na_2EDTA-containing syringes.

2. Centrifuge the blood for 10 min at 1500g, 4 °C, and remove the clear plasma.

3. To a given volume of plasma, add an equal volume of solution (b) from *Table 1* (density 1.117 g/ml) to bring to a final density of 1.063 g/ml. Because of possible run-to-run variations, it is prudent to measure the density and adjust it to 1.063 g/ml if necessary.

4. Place the adjusted plasma in ultracentrifuge tubes and insert the caps. The volume added to each tube should be restricted in order to avoid disturbing the upper layer of plasma when the cap is removed after centrifugation yet a sufficient volume must be present to prevent the tube from collapsing under the high centrifugal force. If the Beckman preparative ultracentrifuge is to be used, add a volume of 10.4 ml to each polyallomer centrifuge tube (16 × 76 mm). If the last tube to be

Protocol 1. *Continued*

  dispensed contains less than 10.4 ml, the volume should be made up to this amount by the addition of solution (a) (density 1.063 g/ml).

5. Check that the tubes are balanced, using solution (a) in any counter-poise tube required.

6. Centrifuge at 40 000 r.p.m. (105 400g) in a fixed-angle rotor (Beckman type 40 or 50Ti) at 16 °C for 18 h. Allow the rotor to decelerate with the brake turned off.

7. Carefully remove the centrifuge tubes, one at a time, from the rotor, take off the caps and collect the upper mobile layer containing the VLDL and LDL along with the upper half of the clear solution above the reddish-coloured layer of HDL and plasma proteins at the bottom of the tube. To achieve this, use either a pasteur pipette or a Beckman tube slicer, in which case the position of the tube is adjusted so that it is sliced at a distance of 3.0 to 4.0 cm from the top.

8. With a fine glass rod, stir the remaining contents (infranatant) of each tube to suspend the small gelatinous pellets that are usually present at the bottom of the tubes and which trap some of the HDL.

9. Pool the infranatant solutions and add an equal volume of solution (f) (density 1.357 g/ml) to bring the final density to 1.21 g/ml, and confirm the latter by measurement in the density meter. (Alternatively, solid NaBr may be added to avoid the use of large volumes.)

10. Place the adjusted solution in centrifuge tubes (as in 4 above), using solution (d) to fill any counterpoise tubes.

11. Centrifuge at 40 000 r.p.m. for 40 h at 16 °C.

12. Remove the whole of the upper layer in a volume of approximately 2.0 ml. If a quantitative recovery is required, collect the upper layer in a volume of about 1.5 ml by slicing the tube at a distance of 2.0 cm from the top, and make the volume up to 2.0 ml with solution (d) in a volumetric flask. Always ensure that the entire HDL layer, extending a few mms into the clear infranatant solution beneath it, is collected to avoid excluding any small, dense, trailing HDL particles. If necessary, increase the volume of HDL collected to make this possible.

2.2 Precipitation methods

The use of precipitation methods for the determination of HDL concentration in plasma is discussed in detail in Chapter 1. While such measurements are of considerable value in clinical investigations, the procedures are of limited application to the analysis of HDL subfractions because the final fractions, though devoid of other lipoprotein classes, contain HDL in an admixture with plasma proteins, and are therefore not a good starting point for the isolation of HDL subfractions. Furthermore, precipitation procedures are no less destructive to the lipoprotein particles than is ultracentrifugation with respect to the dissociation of apolipoprotein (15).

3. The separation and measurement of HDL_2 and HDL_3

3.1 Preparative ultracentrifugation

The method employed is the same as that discussed above in *Protocol 1*, but with inclusion of an additional density step at 1.125 g/ml. The merits of the method are that it yields pure preparations of HDL_2 and HDL_3 that may be useful starting materials for further subfractionation and also that it provides more reliable values than precipitation methods for hypertriglyceridaemic patients.

Protocol 2. The separation of HDL_2 and HDL_3 by flotation

1. Carry out steps 1 to 8 described in *Protocol 1*.

2. Add an equal volume of solution (c) (density 1.187; see *Table 1*) to the pooled infranatant solution to bring to a final density of 1.125 g/ml.

3. Centrifuge at 40 000 r.p.m. for 40 h at 16 °C.

4. Remove the upper layer containing the HDL_2 in a volume of 2 ml.

5. Add an equal volume of solution (e) (density 1.295 g/ml) to the lower layer to give a final density of 1.21 g/ml.

6. Centrifuge at 40 000 r.p.m. for 40 h at 16 °C.

7. Collect the upper layer containing HDL_3 in a volume of 2 ml.

3.2 Precipitation methods

Due to its simplicity, the Gidez–Eder method (16) is used extensively in clinical investigations for the determination of HDL_2 and HDL_3 concentrations in plasma. The basis of the method is that apo B-containing lipoproteins are precipitated from plasma with heparin–$MnCl_2$, leaving the total HDL, along with other plasma proteins, in solution. The HDL_2 fraction is then precipitated from this solution by the further addition of dextran sulphate to leave the HDL_3 fraction in the final supernatant.

Protocol 3. The separation of HDL_2 and HDL_3 by precipitation

Reagents

(a) Heparin: Dissolve 490 mg heparin sodium salt (Sigma) in 2 ml of distilled water.

(b) 1.06 M $MnCl_2$: Dissolve 209.8 g of $MnCl_2.4H_2O$ in distilled water and make the volume up to 1 litre.

(c) Heparin/$MnCl_2$ solution: Add 1.5 ml of solution (a) (see *Table 1*) to 25 ml of solution (b) to give a final within-assay concentration of 1.26 mg of heparin per ml and 0.091 M $MnCl_2$.

(d) Dextran sulphate solution: Dissolve 1.43 g dextran sulphate (mol. wt 15 000; Sochibo) in 0.15 M NaCl and make up to 100 ml with 0.15 M NaCl. It is essential to use the prescribed dextran sulphate to obtain a reproducible separation (16).

Procedure

1. Add 0.3 ml of solution (c) to 3 ml of plasma. Mix thoroughly and allow to stand at room temperature for 10 to 20 min.

2. Centrifuge at 1500g for 1 h at 4 °C. Remove a portion of the clear supernatant immediately for step 3 of the procedure and for analysis of total HDL cholesterol.

3. To 2 ml of the supernatant add 0.2 ml of solution (d). Mix thoroughly and let stand at room temperature for 20 min.

4. Centrifuge at 1500g for 30 min at 4 °C and remove an aliquot of the clear supernatant immediately for analysis of HDL_3 cholesterol.

5. Measure the cholesterol concentration of total HDL (step 2) and HDL_3 (step 4) (see Section 5.4.1). HDL_2 cholesterol = total HDL cholesterol minus HDL_3 cholesterol. Note that this method does not yield HDL_2 or HDL_3 free of plasma proteins.

4. The separation of HDL subfractions

4.1 Gradient gel electrophoresis

4.1.1 Principle

The general principles underlying the use of gradient gel electrophoresis are described in Chapter 1 and an excellent account of gel electrophoresis is to be found in reference 17. While the technique is capable of resolving most protein mixtures into sharp, discrete bands, HDL produces broad and overlapping zones due to the high degree of heterogeneity, with subspecies of similar and overlapping size. Nevertheless, the resulting profile is highly reproducible and provides a reliable quantitative measure of the distribution of HDL subpopulations, provided the protocol described below is rigidly adhered to (11).

4.1.2 Polyacrylamide gradient gels

Normally gels with a 4–30% linear gradient of acrylamide are employed for the separation of HDL subfractions, though concave gels are particularly useful when LDL or other lipoproteins are also present (18). Commercially manufactured gels (e.g. Flowgen—most previous analyses have been made with Pharmacia gels, but production of those has recently stopped), though expensive, provide the greatest convenience. They have, however, been reported to be unsuitable for studies where the resolved proteins are subsequently transferred to nitrocellulose for immunological examination, because of poor transfer efficiency. For such purposes, the investigator should produce his own gradient gels (19). Laboratory-made gels, however, usually lack the precision and reproducibility of manufactured gels and may be less suitable for quantitative studies. The best means of making gels is by the use of gel-casting kits (Pharmacia/LKB; BioRad; see reference 19) though laboratory-made gradient devices may produce satisfactory results.

4.1.3 Electrophoretic separation

i. *Reagents*

(a) 0.09 M Tris/0.08 M boric acid/3 mM EDTA (pH 8.35) (reservoir buffer): dissolve 10.90 g of Tris, 4.95 g of H_3BO_3, and 1.12 g of Na_2EDTA in distilled water, adjust to pH 8.35, and make up to 1 litre.

(b) 5-Sulphosalicylic acid (10%): Dissolve 50 g 5-sulphosalicylic acid in distilled water and make the volume up to 500 ml.

(c) Coomassie blue stain: Dissolve 0.01 g Coomassie brilliant blue R250 in methanol:acetic acid:water (5:1:4, v/v/v) and make up to 1 litre with this mixture.

ii. *Sample preparation*

For the analysis of total HDL, use the material isolated by ultracentrifugation at density 1.21 g/ml (Section 2.1) directly and without dialysis. When isolated subfractions of HDL are to be examined, dialyse the samples (after concentration, if necessary) against buffer (a) and mix four parts of this solution with one part of a solution containing 40% sucrose in buffer (a) (to aid in the layering of the sample on top of the acrylamide gel). Samples of HDL or its subfractions may be stored for up to 24 h at 4 °C.

Protocol 4. Method for the separation of lipoproteins by electrophoresis

1. Fit the filled glass sandwich into the electrophoresis tank.

Protocol 4. *Continued*

2. Place a spacer comb tightly and in a level position on top of the gel with sufficient pressure to ensure that no channel remains between the bottom edge of the spacer and the gel that would allow samples to diffuse between adjacent wells. At the same time, avoid tearing or cracking the gel.

3. Add the appropriate volume of buffer (a) to the tank. Remove any air bubbles that may be trapped in the wells by forcing a jet of buffer into each well from a pasteur pipette.

4. Pre-equilibrate the gels for 20 min at 70 V.

5. Turn off the current. By means of an automatic pipette, carefully load 10 to 20 µl of sample (containing 15 to 20 µg of protein) into alternate wells. To two of the wells apply a similar volume of a reference protein mixture for calibration of particle size radius (e.g. HMW Calibration Kit, Pharmacia, containing thyroglobulin, apoferritin, catalase lactic dehydrogenase, and bovine serum albumin—see Chapter 1). To avoid streaking of lipoprotein bands during electrophoresis, some investigators recommend applying to the intervening wells the same volumes of a mixture containing four parts of a solution with a salt concentration equivalent to that of the lipoprotein sample and one part of 40% sucrose in this buffer; this procedure has not, however, been found to be of value in the author's laboratory.

6. Carry out the electrophoresis at 20 V for 20 min, then 70 V for 30 min, and finally 120 V for 24 h.

7. Turn off the current. Remove the gel sandwich from the tank, separate the glass plates and fix the gels in solution (b) for 1 h immediately following electrophoresis. Then stain with Coomassie blue (solution (c)) for 1.5 h and destain the gel with methanol:acetic acid:water (5:7.5:87.5, v/v/v) for several days with gentle horizontal rotation and frequent change of wash solution. Store the gels in 9% acetic acid in which the stained bands are stable for at least 3 months at room temperature.

4.1.4 Densitometry

While visual observation with photographic recording (*Figure 1*) of the resolved HDL subspecies is satisfactory for many purposes, the complex and overlapping nature of the bands necessitates the requirement for scanning and measurement of band areas for a full evaluation of the results to be appreciated. Several commercially available instruments are suitable for this purpose (BioRad Model 620 video densitometer, Pharmacia/LKB Ultrascan XL Laser densitometer, Beckman Appraise, Joyce Loebl Chromoscan).

Gradient gel electrophoresis resolves HDL from human subjects into five

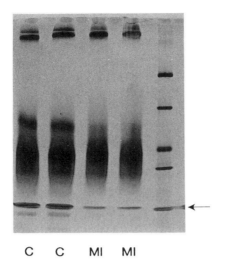

C   C   MI   MI

Figure 1. Gradient gel electrophoresis (4–30% polyacrylamide) showing the separation of subfractions of HDL from two healthy, male control subjects (C) and from two male survivors of myocardial infraction (MI). The bands were stained with Coomassie Blue. Arrow indicates serum albumin.

subfractions which fall within the R_F ranges shown in *Table 2* (see *Figure 2*). Two of the subfractions (HDL_{2a} and HDL_{2b}) correspond to HDL_2 and three of the components (HDL_{3a}, HDL_{3b}, and HDL_{3c}) correspond to HDL_3 isolated by ultracentrifugation. The mean particle diameters of the subfraction obtained by this method correspond to those measured by electron microscopic methods and determinations made in the analytical ultracentrifuge (20, 21). The method provides the basis of a useful system for defining HDL subfractions and therefore the distribution of subfractions within the R_F ranges shown in *Table 2* should be determined by measurement of peak areas within these limits. For this purpose, the R_F value is defined as the migration distance of a peak of a particular HDL subfraction relative to the migration distance of the peak of bovine serum albumin, included as a standard (11).

Table 2. Designation of HDL subfractions separated by gradient gel electrophoresis[a]

HDL subfraction	R_F range	Mean diameter (nm)
HDL_{3c}	0.841−0.962	7.62
HDL_{3b}	0.781−0.841	7.97
HDL_{3a}	0.711−0.781	8.44
HDL_{2a}	0.627−0.711	9.16
HDL_{2b}	0.445−0.627	10.57

[a] Data taken from reference 11.

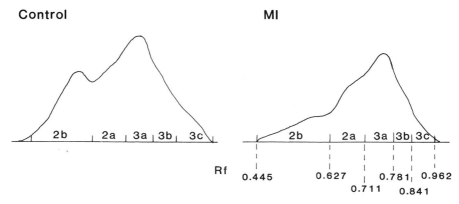

Figure 2. Densitometric scans of HDL subfractions from a control subject and a survivor of myocardial infraction shown in *Figure 1*, obtained with the Bio-Rad Model 620 video densitometer.

Profound differences occur in the distribution of these HDL subfractions in different individuals, though most subjects maintain a fairly uniform profile provided that they adhere to a constant life-style. Significant changes in the levels of specific subfractions occur with certain forms of physical exercise, dietary modification, alochol intake, and the administration of some drugs (22, 23). Subjects with a high coronary risk (24) and survivors of myocardial infraction (H. M. Wilson, and E. R. Skinner, unpublished data) have significantly lower levels of HDL_{2b} than control subjects. The HDL of several experimental animals (e.g. rabbit, rat) is resolved by gradient gel electrophoresis into subfractions with a profile that resembles that of man, though the subfractions are often of larger particle size than their human counterparts. It should not be overlooked that the resolved subfractions are heterogeneous and contain HDL subspecies of similar size, yet of different composition and possibly serving different metabolic roles. Changes in R_F maxima within the R_F ranges indicated in *Table 2* may therefore be important.

4.2 Heparin–Sepharose affinity chromatography

4.2.1 Principle

The observation (25) that heparin–Sepharose binds a subfraction of HDL that contains apo E has stimulated the fairly wide use of this method. Since apo E-rich HDL interacts with tissue receptors and has been reported to be implicated in reverse cholesterol transport, the method provides a separation that may correlate with metabolic parameters. Apo E-rich HDL separated by this technique is heterogeneous with respect to its content of apo E and recent attempts (26) have been made to fractionate the different apo E rich subspecies by the use of a series of eluting solvents of different ionic composition in order that the metabolic role of these components may be investigated.

4.2.2 Preparation of heparin–Sepharose

Coupled heparin–Sepharose is available commercially or may be produced in the laboratory from activated Sepharose, following the manufacturer's instructions (Pharmacia/LKB). Some investigators have found that these products do not always give satisfactory and reproducible results. In the author's experience, the full laboratory preparation from heparin and Sepharose, though rather laborious, yields a product of high binding capacity which gives quantitatively reproducible results. The method is based on the procedures described in references 27 and 28.

i. Reagents

(a) 2 M Na_2CO_3: Dissolve 21.2 g of Na_2CO_3 in distilled water and make up to 100 ml.

(b) Cyanogen bromide/acetonitrile: Dissolve 10 g CNBr (fresh) in 5.0 ml of acetonitrile. Make up immediately before use and stand on ice.

(c) 0.2 M sodium bicarbonate (pH 9.5): Dissolve 25.2 g of $NaHCO_3$ in distilled water, adjust the pH to 9.5 with NaOH and make up to 3 litres.

(d) 0.1 M Sodium bicarbonate (pH 9.5): To 1.5 litres of solution (c), add 1.5 litres of distilled water.

(e) Heparin/sodium bicarbonate: Dissolve 0.5 g heparin (sodium salt) in solution (c) and make up to 100 ml with this solution. The molecular weight and composition of heparin varies according to source. Heparin (sodium salt) from porcine intestinal mucosa (Sigma) has been found to give reliable results using the above method.

(f) 0.1 M Sodium acetate buffer (pH 4.0): Dissolve 16.40 g of sodium acetate in distilled water, adjust to pH 4.0 with acetic acid and make the volume to 2 litres with distilled water.

(g) 2.0 M urea: Dissolve 240 g of urea in distilled water and make up to 2 litres.

(h) 2 M NaCl: Dissolve 116.9 g NaCl in distilled water and make up to 1 litre.

(i) 5 mM Tris–HCl/0.05 M NaCl/0.03 M thiomersal (pH 7.4): Dissolve 0.2 g of thiomersal in 2 litres of 5 mM Tris–HCl/0.05 M NaCl (see reagent (a) in Section 4.2.3 below).

Protocol 5. The preparation of heparin–Sepharose

The preparation should be carried out in a fume cupboard in a well-ventilated room because cyanogen bromide emits poisonous and irritating cyanide fumes.

Protocol 5. *Continued*

1. Dilute a packed volume of 50 ml of Sepharose 4B (Pharmacia/LKB) to 100 ml with distilled water.

2. Add 100 ml of solution (a) (2 M Na_2CO_3) slowly with stirring.

3. With the solutions maintained on ice, add the 5 ml volume of solution (b) to the slurry in one lot and stir vigorously for 2 min.

4. Pour the slurry onto a sintered glass funnel and wash with 1 litre of solution (d), followed by 1 litre of distilled water, and finally 1 litre of solution (c). Filter under vacuum to give a moist compact cake.

5. Transfer this material to a *plastic* bottle containing 100 ml of solution (e) (heparin/$NaHCO_3$). Mix and allow to stand at 4 °C for 20 h.

6. Add 7.5 g of glycine to the mixture to mask any uncoupled reactive groups. Mix and stand for 18 h at 4 °C.

7. Wash the heparin–Sepharose on a coarse sintered-glass filter with 2 litres of each of the following solutions consecutively: *i.* solution (f); *ii.* solution (g); *iii.* solution (d); *iv.* solution (h); *v.* solution (i).

8. Filter under vacuum to give a moist, firm material. Transfer this to a stoppered container and add an equal volume of solution (i). This final product is stable for at least 6 months at 4 °C.

4.2.3 Chromatography

i. Reagents

(a) 5 mM Tris–HCl/0.05 M NaCl (pH 7.4): Dissolve 1.83 g of Tris and 8.76 g of NaCl in distilled water. Adjust the pH to 7.4 with HCl and make up to 3 litres.

(b) 5 mM Tris–HCl/0.05 M NaCl/0.025 M $MnCl_2$ (pH 7.4): Dissolve 0.61 g of Tris and 2.92 g of NaCl in distilled water, adjust to pH 7.4 with HCl and make up to 1 litre. *Then* add 4.95 g of $MnCl_2$. (Add the $MnCl_2$ immediately before use as $MnCl_2$ precipitates out on prolonged standing.)

(c) 5 mM Tris–HCl/0.1 M NaCl (pH 7.4): Dissolve 0.61 g of Tris and 5.84 g of NaCl in distilled water, adjust to pH 7.4 with HCl and make to 1 litre.

(d) 5 mM Tris–HCl/0.6 M NaCl (pH 7.4): Dissolve 0.61 g of Tris and 35.1 g of NaCl in distilled water. Adjust to pH 7.4 with HCl and make to 1 litre.

ii. Sample preparation

(a) Dialyse the sample of HDL (prepared by ultracentrifugation in the density range 1.063 to 1.21 g/ml; Section 2.1) against 50 volumes of solution (a) with stirring and at 4 °C for at least 18 h with several changes of buffer.

(b) Adjust the concentration to give 4–8 mg of HDL protein/ml by appropriate addition of buffer (a).

(c) Just before applying to the column, add solid $MnCl_2$ to the HDL solution to give a final Mn^{2+} concentration of 25 mM. (i.e. add 5.0 mg $MnCl_2$ per ml).

Protocol 6. Separation of apo E-rich and apo E-poor HDL by heparin–Sepharose affinity chromatography

1. Stir the heparin–Sepharose slurry and allow to settle for a few minutes before removing the supernatant to eliminate any 'fines'. Add an equal volume of solution (a). Stir and degas the slurry.

2. Pack the slurry into a column (12 cm × 1.0 cm i.d.), preferably with an adjustable capacity adaptor to eliminate dead-space at the top of the column.

3. Carry out the remainder of the procedure at 4 °C. Allow buffer (a) to flow through the column for at least 18 h at a flow rate of 12–15 ml/h to ensure that equilibrium has been reached.

4. Immediately before applying the sample, pass a volume of 20–30 ml of solution (b) through the column.

5. Apply 1.5 ml of the HDL solution (to which $MnCl_2$ has been added: see above) to the column. Wash in with 2–3 ml of buffer (b) and leave the HDL in the column overnight to permit maximum binding.

6. With the column attached to a recorder and fraction collector, elute the non-binding proteins (Fraction I) with buffer (b) until the absorption at 280 nm falls to the baseline value (about 20 ml). Collect elution volumes of 4 ml.

7. Replace the eluting buffer, solution (b), with buffer (c) and continue to elute with this buffer until the bound HDL (Fraction II) is completely removed from the column.

8. Finally elute with buffer (d) and collect Fraction III.

9. Pass at least 30 ml of buffer (a) through the column to equilibrate it in readiness for the next chromatographic separation.

10. If quantitative results are required, measure the absorption of the column fractions at 280 nm in a UV spectrophotometer.

A typical elution diagram is shown in *Figure 3*. Fraction I (non-bound HDL) contains HDL subfractions that are largely devoid of apo E and are mostly confined to the HDL_3 subclass, though some HDL_2 components are also present in this fraction (*Figure 4*). In Fraction II (bound HDL) are the apo E-rich HDL subfractions; these are mainly of large particle size which fall

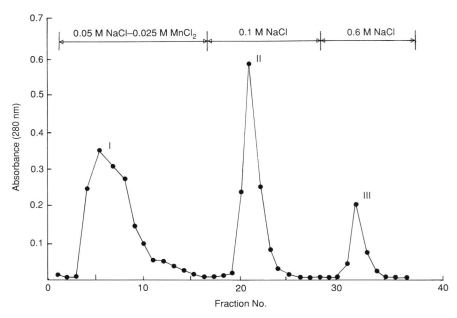

Figure 3. Heparin–Sepharose affinity chromatography of HDL. I: non-bound, apo E-poor HDL; II; bound apo E-rich HDL; III: Lp(a) and LDL.

within the HDL_2 subclass. Fraction III contains small quantities of LDL and Lp(a). The individual eluted fractions that comprise Fraction I and those that form Fraction II may be pooled and concentrated for further analytical or metabolic studies. Stirred cells (Amicon, Sartorius) or centrifugal concentration tubes such as the Centriprep or Centrisart 1 (Amicon, Sartorius) are very suitable for this purpose. Reference 29 provides a useful account of the methods available for concentrating protein solutions.

Plasma concentration of apo E-poor and apo-E rich HDL are calculated from a summation of the A_{280} values of the individually collected fractions that make up Fraction I and Fraction II and applying a conversion factor (0.41 mg of HDL-protein/ml per A_{280} unit). The method gives good reproducibility; the coefficient of variation for the estimate of the concentration of apo E-rich HDL on a single subject measured on five occasions over a two-year period was 9.5% (24).

4.3 Immunoaffinity chromatography

4.3.1 Principle

Immunoaffinity chromatography provides a valuable tool with considerable potential for the separation of HDL subfractions because it is one of the few

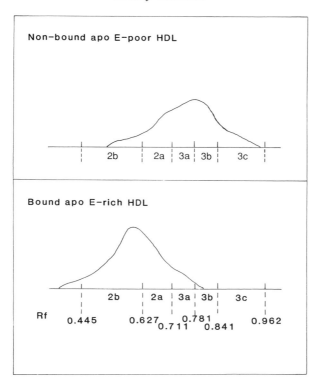

Figure 4. Gradient gel densitometric scans of HDL subfractions separated by heparin–Sepharose affinity chromatography obtained with the Bio-Rad Model 620 video densitometer.

methods that avoids the use of ultracentrifugation in high salt concentrations, a treatment which causes appreciable dissociation of some apolipoproteins from the native HDL particle. Plasma is mixed with immobolized mono-specific antibody and, after removal of non-bound and non-specifically bound proteins and lipoproteins by washing, the complexed lipoprotein species containing the specific apolipoprotein antigen is removed with appropriate solvents. The method has been used (30) for the isolation of HDL particles which contains both apo AI and AII (AI-with-AII) and particles which contains apo AI but no apo AII (AI-without-AII). It has also provided a means of separating apo E-containing lipoproteins (31). Valuable information on the physical characteristics and the metabolic roles of HDL subfractions of specific apoprotein composition has been gathered by this technique (32, 33). Several forms of the method have been successfully employed (33, 34) including the procedure described below for the separation of apo AI and apo AII-containing HDL (30). It must be emphasized, however, that antibodies differ considerably with regard to affinity and other characteristics, so that the conditions required for both binding and elution may need modification with different preparations of antibody.

4.3.2 General method

i. Reagents

(a) 0.2 M sodium phosphate/0.5 M NaCl (pH 6.8):

    i.   Dissolve 31.2 g of $NaH_2PO_4.2H_2O$ in distilled water and make up to 1 litre.

    ii. Dissolve 71.6 g of $Na_2HPO_4.12H_2O$ in distilled water and make up to 1 litre.

    Mix 51 parts of solution (i) to 49 parts of solution (ii), to give a pH of 6.8 and add 29.2 g NaCl per litre of solution.

(b) 0.01 M Tris–HCl/0.15 M NaCl/1 mM EDTA/0.02% thiomersal (pH 7.4): Dissolve 1.21 g Tris in distilled water and adjust the pH to 7.4 with HCl. Add 8.77 g of NaCl, 0.37 g of Na_2 EDTA, and 0.2 g thiomersal and make the volume up to 1 litre.

(c) 0.1 M $NaHCO_3$/0.5 M NaCl/1 mM EDTA (pH 8.0): Dissolve 8.40 g of $NaHCO_3$ in distilled water and adjust the pH to 8.0. Add 29.2 g of NaCl and 0.37 g of Na_2EDTA and make up to 1 litre.

(d) 0.5 M acetic acid/0.5 M NaCl/1 mM EDTA (pH 3.0): Dissolve 29.2 g of NaCl and 0.37 g Na_2EDTA in 0.5 M acetic acid and make up to 1 litre in this solution.

(e) 3 M Sodium thiocyanate: Dissolve 24.3 g of NaSCN in distilled water and make the volume up to 1 litre.

Protocol 7. General method for immunoaffinity chromatography

1. Conjugate antibodies monospecific for apo AI and for apo AII to Sepharose 4B at a ratio of 7–8 mg immunoglobulin per ml of gel in solution (a). Details of the method for conjugation, as well as for the production of antibody, are to be found in Chapter 3.

2. Store the immunosorbent in buffer (a) at 4 °C until required.

3. Mix 1 ml of plasma with 2 ml of immunosorbent for 1 h at 4 °C. The amount of immunosorbent needed depends on its binding capacity and it is advisable to establish this in a preliminary experiment.

4. Pack the HDL–immunosorbent complex into a borosilicate column of approximately 1.5 cm i.d.

5. Wash the non-linked proteins from the column with buffer (b).

6. Remove non-specifically bound proteins with solution (c).

7. Finally elute apo AI- or apo AII-containing lipoprotein particles with either solution (d) (and immediately add solid Tris to buffer the solution) or elute with solution (e). Both eluants provide a good recovery of HDL. It is reported that the recovery of lecithin:cholesterol acyltransferase is higher when acetic acid is employed but cholesteryl ester transfer activity can only be detected when thiocyanate is used as the desorbing agent (30).

8. Concentrate the eluted samples to appropriate volumes for further studies.

4.3.3 The isolation of AI-with-AII lipoproteins

Incubate fresh plasma with anti-apo AII immunosorbent for 1 h at 4 °C, pack into a small borosilicate column and sequentially wash the nonbinding proteins, nonspecifically-bound protein, and apo AII-containing lipoproteins from the column as detailed above.

4.3.4 The isolation of AI-without-AII

Incubate the nonbinding plasma proteins obtained in Section 4.3.3 above with anti-apo AI immunosorbent for 1 h at 4 °C and elute the nonbinding proteins, nonspecifically bound proteins and the AI-without-AII lipoproteins as described.

4.4 Other methods for separating HDL subfractions

Gel filtration chromatography has been used (35) to isolate subfractions of HDL in sufficient quantity for their compositions to be determined. The resolving power of the method is insufficient to allow the separation of HDL subfractions into discrete bands and so the HDL elutes as a single, broad peak. However, fractions of the elute containing essentially single populations of HDL_{2b}, HDL_{2a}, HDL_{3a}, and HDL_{3b} may be identified by gradient gel electrophoresis (see Section 4.1).

Although methods based on adsorption chromatography have not attained general use, hydroxylapatite has been employed with some success, at least in combination with other methods, to separate HDL subfractions which differ in their apolipoprotein composition (36). The method consists of applying HDL (or HDL_2 or HDL_3), prepared by ultracentrifugation, to a column containing hydroxylapatite and eluting by stepwise addition of potassium phosphate buffers, ranging from 0.05 M to 0.65 M at pH 6.8.

Pevikon block electrophoresis has been employed for the isolation of HDL_c, a subclass of HDL which is enriched in apo E. It is induced in several animal species by cholesterol feeding and may correspond to HDL_1 in man. The procedure involves electrophoretic separation of HDL_c from other lipoprotein species on a flat bed containing Pevikon beads as support medium and the subsequent elution of this component from the region of the bed volume to which it has migrated. The interested reader should consult reference 37 for details.

5. The analysis and characterization of HDL subfractions

The effectiveness of the above procedures in achieving a fractionation of the many subspecies of which HDL is composed should be assessed by evaluating the mean particle size and the composition of the subfractions that have been isolated. A detailed knowledge of these parameters, particularly of apolipo-protein composition, is of paramount importance in understanding the underlying mechanisms when the purpose of the separation is to investigate the metabolic roles of the subfractions or to determine alterations in their concentrations in clinical disorders or in response to physiological, dietary, therapeutic or other stimuli.

5.1 The analysis of intact HDL particles

5.1.1 Gradient gel electrophoresis

This method (see Section 4.1) affords a relatively simple and rapid means of gaining an initial assessment of the degree of fractionation achieved before more detailed analysis of subfraction composition is undertaken. Because the distance migrated along the gel is a function of particle size, it should not necessarily be assumed that subfractions that fall within the same R_F region (see *Table 2*) are identical since they may represent species that have similar sizes but are of different compositions. Very small differences in R_F may be present in such cases and it is therefore prudent to make the comparison between subfractions on a single gel or, where this is not possible, to include a reference standard (in the form of total HDL from a single individual) in each gel employed. This will also eliminate errors due to possible gel-to-gel variation in the degree of cross-linkings. The use of a sensitive densitometer is essential as visual observation may fail to detect small differences in R_F.

5.1.2 Electron microscopy

Electron microscopy provides a valuable technique for gaining an insight into the size distribution and morphology of negatively stained HDL and its subfractions. In essence, HDL, prepared by ultracentrifugation, or its subfractions isolated by the procedures described above, are dialysed against an appropriate buffer, adjusted to a concentration of 100 to 250 µg of lipoprotein protein/ml in 1% sodium phosphotungstate and added as a small droplet to a Formvar-carbon coated grid for examination at a magnification of 40 000 to 60 000. The method requires expensive and very exacting technique and the reader is referred elsewhere for full details (38).

In addition to providing important information on particle dimensions, electron microscopy is of considerable value for the detection of atypical, non-spherical HDL such as discoidal nascent HDL observed in liver perfusates (39) and square-packed structures in the less dense HDL fraction of lymphedema fluid (40). Caution, however, has to be exercised against the production of artefacts, including the formation of clumps and aggregates, and their tendency to lose content during processing. The mean particle diameters of HDL subfractions obtained by electron microscopy confirm those obtained by gradient gel electrophoresis and analytical ultracentrifugation (11,20).

5.1.3 Analytical ultracentrifugation

This method, which also requires expensive equipment and experience in operation, produces useful information on HDL particle size and forms a useful adjunct to electron microscopy and gradient gel electrophoresis. HDL subfractions are dialysed against NaBr/NaCl solutions of density 1.200 at 26 °C and the corrected flotation rates at this density ($F\,°_{1.20}$) are calculated from the position of the schlieren peaks at time intervals during ultracentrifugation. Studies in the Donner Laboratory (11, 20, 21) have demonstrated that the particle size diameter of three HDL components identified by analytical ultracentrifugation are close to those of HDL_{2b}, HDL_{2a}, and HDL_3 as separated by gradient gel electrophoresis. The correspondence between the results of ultracentrifugal and gradient gel electrophoresis methods was evidenced by strong correlations between the plasma concentrations of these components in a large population as determined by analytical ultracentrifugation and normalized scan areas on the gradient gels.

5.2 Determination of apolipoprotein composition by electrophoresis

5.2.1 Delipidization

The apolipoprotein composition of HDL and its subfractions may be determined by submitting them directly to electrophoresis in appropriate buffers containing dissociating agents such as SDS or urea. The resulting apolipoprotein bands are often poorly resolved as a result of interference by the lipid moieties and it is therefore difficult to obtain reliable quantitative data by this method. It is therefore strongly recommended to delipidize lipoprotein samples before carrying out the electrophoretic separation. Of the many delipidization procedures that have been reported (41), two methods have gained general acceptance and are used widely for routine analysis in many laboratories; the loss of small apolipoproteins (such as apo C) due to their solubility in organic solvents is minimal in these procedures.

Protocol 8. Delipidization of HDL by precipitation with ethanol—ether (42)

Reagent

0.15 M NaCl/1 mM EDTA: Dissolve 17.5 g NaCl and 0.74 g Na_2EDTA in distilled water. Adjust the pH to 8.0 and make up the volume to 2 litres with distilled water.

Procedure

1. Dialyse the sample of HDL or its subfractions (containing about 5 mg protein/ml) against the above solution overnight at 4 °C with several changes of buffer.
   The following procedure should be carried out in a laboratory free from electric sparks.

2. Rapidly inject the dialysed solution into 50 volumes of ethanol:diethyl ether (3:2,v/v) that has been pre-cooled to −20 °C in a glass-stoppered flask. Use a syringe or automatic pipette to ensure that the injected solution is dispersed into small droplets to provide a large surface area of contact with the organic phase.

3. Stand the mixture for 90 min at −20 °C **in a spark-free refrigerator** with occasional shaking.

4. Transfer the suspension to a pre-cooled centrifuge tube and centrifuge for 10 min at 700g in a centrifuge that is **spark free and has been cooled to below −20 °C. Extreme precaution must be taken to avoid explosion caused by ether whose flash point is below −20 °C; the temperature must not be allowed to rise above −20 °C.** Also check that the bearings are not allowed to warm up and fix a suitable cap on the centrifuge tube. Screw-capped centrifuge tubes should be used to prevent the spread of ether vapour. Tubes must be solvent-proof and withstand the effect of centrifugation at −20 °C in the presence of ether/ethanol mixtures. In the author's laboratory Teflon FEP tubes (Nalge Co) have been used successfully for this purpose.

5. Carefully decant the supernatant, leaving a small pellet of apolipoprotein in a few drops of solvent at the bottom of the tube.

6. Add a volume of diethyl ether (at −20 °C) equal to that of the original volume and resuspend the pellet by gentle teasing with a fine glass rod.

7. Centrifuge for 10 min at 700g at −20 °C.

8. Decant the supernatant.

9. Pool the first two supernatants (i.e. at steps 5 and 8) and recentrifuge to generate a further pellet of apolipoproteins. Combine this with the original pellet (from step 5).

10. Wash and centrifuge the combined pellet a further five times with half of the original volume of diethyl ether.

11. With a stream of nitrogen, remove the ether remaining at the bottom of the tube after decanting off the supernatant following the last wash. Store the apolipoproteins at −20 °C.

Protocol 9. Delipidization of HDL in the soluble state by ethanol−diethyl ether (43)

Reagent

20 mM Tris–HCl/0.1 M NaCl (pH 8.5): Dissolve 2.4 g Tris in distilled water and adjust the pH to 8.5 with NaOH. Add and dissolve 5.8 g NaCl and make up the volume to 1 litre with distilled water.

Procedure

1. Dialyse the sample of HDL (containing about 4 mg protein/ml) against several changes of the above buffer overnight with stirring at 4 °C.

2. Add an equal volume of diethyl ether/ethanol (3:2, v/v) in a glass-stopped tube at room temperatures. Shake well and stand until the two phases are separated (about 20 min).

3. Remove the top layer with a pasteur pipette.

4. Add the same volume of diethyl ether/ethanol (3:1, v/v), shake, stand, and remove the top layer.

5. Repeat this extraction process five times.

6. Blow off the ether from the final bottom layer to provide an aqueous solution of apolipoproteins. Store this solution at 4 °C for 2 days or at −20 °C for up to 1 month.

5.2.2 SDS Gel electrophoresis

In this procedure, the protein is denatured by heating with SDS in the presence of mercaptoethanol to cleave disulphide bonds. The denatured polypeptides bind SDS in a constant weight ratio to produce SDS–polypeptide complexes which have uniform charge densities and therefore migrate in a polyacrylamide gel at a rate that is a function of molecular size (44). The system described is based on the method of Stephens (45).

i. *Reagents*

(a) 0.25 M Tris glycine (pH 8.3) (gel buffer): Dissolve 60.6 g Tris and 288 g glycine in distilled water; if necessary adjust the pH to 8.3 and make up to a final volume of 2 litres with distilled water.

(b) 25 mM Tris glycine/1 mM EDTA/1% SDS/5% mercaptoethanol (pH 8.3) (sample buffer): Dissolve 37.2 mg Na_2EDTA, 1 g SDS, and 4.5 ml mercaptoethanol in buffer (a) that has been diluted 10 times with water and make up to a final volume of 100 ml with this diluted buffer.

(c) 25 mM Tris glycine/0.05% bromophenol blue/1% sucrose (pH 8.3) (tracking dye). Dissolve 3.03 g Tris, 14.4 g glycine, 50 mg bromophenol blue, and 1 g sucrose in distilled water and make the volume up to 100 ml.

(d) Bis-acrylamide (30:0.8 wt/wt): Dissolve 30 g acrylamide and 0.8 g N,N'-methylenebis-acrylamide in gel buffer (a) and make to a final volume of 100 ml in this buffer.

(e) 10% SDS (stock): Dissolve 10 g sodium dodecyl sulphate in a final volume of 100 ml of gel buffer.

(f) 10% ammonium persulphate: Dissolve 10 g ammonium persulphate in distilled water without heating and make to a final volume of 100 ml. This solution must be made up fresh immediately before mixing the gel components.

(g) 0.25 M Tris glycine/1.0% SDS (pH 8.3) (reservoir buffer stock): Dissolve 10.0 g of SDS in buffer (a) and make up to 1 litre with this buffer. Store at 4 °C. (N.B. Dilute this stock buffer 10 times with distilled water before use.)

The acrylamide mixtures shown in *Table 3* should be prepared immediately before use. Acrylamide is a neurotoxin: rubber gloves should be worn and laboratory safety precautions observed.

Table 3. Composition of acrylamide gel mixes

Stock solution	Final acrylamide concentration in gel	
	4%[a]	10%[a]
Bis-acrylamide (d)	13.3	33.3
0.25 M Tris–glycine (a)	10.0	10.0
10% SDS stock (e)	1.0	1.0
TEMED[c]	0.05	0.05
Distilled water	74.6	55.2
10% Ammonium persulphate (f)[b]	1.0	0.5

[a] Figures in the columns represent the volumes of the reagents required to give 100 ml of mixture.
[b] This reagent should be added last and immediately before pouring the gel.
[c] N,N,N',N'-tetramethylethylenediamine

Protocol 10. Preparation of apolipoprotein sample for SDS electrophoresis

1. Dissolve a portion of the apolipoprotein in buffer (b) containing SDS and mercaptoethanol to give a concentration of 20–60 µg protein per 100 µl of buffer. This should be carried out in small capped microcentrifuge tubes. The scale will depend on the amount of material available, but volumes as small as 50 µl may be used. It is always essential that the weight ratio of SDS to protein be at least 3:1. If the apolipoprotein has been prepared by delipidization in the soluble state, it should be dialysed against buffer (a).

2. Place the tubes in a water bath at 100 °C and incubate for 2 min.

3. Allow the tubes to cool at room temperature.

4. Add 15 µl of mercaptoethanol and 10 µl of solution (c) (tracking dye) per 100 µl of apolipoprotein solution.

5. Apply a volume of 20 µl to each sample well in the acrylamide slab (see below).

Variations of this procedure (44) may be performed to establish that complete denaturation of the protein has been achieved and that small mol.wt polypeptides do not represent products of proteolysis.

The electrophoresis is best carried out on slabs using a commercially-available system (Pharmacia/LKB, BioRad, Gibco). Since the procedure differs somewhat with systems from different manufacturers, only an outline description will be given and the manufacturers instructions should be followed. Apolipoproteins may also be resolved by tube gel electrophoresis (see Section 5.2.3), but unless this forms the first dimension of a two-dimensional separation, its use is not recommended because of the ensuing problems that arise in comparing samples and in the scanning of the rod-shaped gels.

Protocol 11. Procedure for SDS electrophoresis

1. Two glass plates (usually 16 × 16 cm and 16 × 20 cm) are sealed between spacers on two sides and on the bottom with 4% acrylamide mix (see *Table 3*) to the prescribed level.

2. Fill the space between the glass plates with 10% acrylamide mix (*Table 3*), overlay with distilled water, and allow to polymerize at room temperature (about 30 min).

3. Remove the water layer, apply a layer of 4% acrylamide mix on top of the formed gel and insert the spacer comb immediately. Leave undisturbed to polymerize. If desired, the polymerized gel may be wrapped in film and stored at 4 °C overnight.

4. Insert the gel sandwich into the electrophoresis tank and fill the latter with buffer (one volume of reservoir buffer stock (solution (g)) and nine volumes of distilled water).

Protocol 11. *Continued*

5. Apply a volume of 200 μl of apolipoprotein sample to each well using an automatic pipette. Also apply a similar volume of marker proteins to two of the wells for molecular weight calibration. Suitable mixtures for this purpose are the 'Electran' mol. wt. range 12 300 to 78 000 (BDH) and the low molecular weight range (14 000 to 97 400) (BioRad) standards.

6. Electrophorese at a constant current of 24 mA until the tracking dye has migrated to within a few cm of the bottom of the gel (about 5 h).

7. Turn off the current. Remove the gel chamber from the electrophoresis tank and carefully separate the gel from the glass plates.

8. Fix and stain the gel as described in Section 4.1.3.

5.2.3 Electrophoresis in 8 M urea

In this form of electrophoresis, the rate of migration of apolipoproteins down the gel is determined by net charge as well as by molecular size. It is particularly useful for identifying the different apo C proteins and is conveniently carried out in gel tubes (46) as described below, or it may be performed on a slab.

i. Reagents

(a) 3.0 M Tris–HCl (pH 8.9) (gel buffer): Dissolve 36.6 g Tris in distilled water, adjust the pH to 8.9 with HCl and make to a final volume of 100 ml.

(b) Bis-acrylamide: Dissolve 30 g acrylamide and 0.8 g N,N'-methylenebis-acrylamide in distilled water and make to a volume of 100 ml.

(c) Ammonium persulphate: Dissolve 0.14 g ammonium persulphate in distilled water and make the volume to 100 ml.

(d) 0.05 M Tris glycine (pH 8.3) (reservoir buffer stock): Dissolve 6.0 g Tris and 28.8 g glycine in distilled water and make the volume to 1 litre. (N.B. Dilute 10 times before use.)

(e) 50 mM Tris glycine/0.05% bromophenol blue/1% sucrose (pH 8.3) (tracking dye): Dissolve 50 mg bromophenol blue and 1 g sucrose in 100 ml of reservoir stock buffer which has been diluted 10 times.

(f) Gel mixture: Immediately before use, rapidly dissolve 7.7 g of urea in a mixture containing 2 ml of solution (a), 4 ml of solution (b), 2 ml of distilled water, 5 μl of TEMED (N,N,N'-tetramethylethylenediamine), and 8 ml of solution (c).

ii. Apolipoprotein sample preparation

Dissolve the apolipoprotein in solution (e) (tracking dye), to which 0.48 g urea per ml has been added, to give a solution containing between 20 and 60 μg protein per 100 μl of solution.

Protocol 12. Procedure for electrophoresis in 8 M urea

The electrophoresis is usually performed with a commercially-available kit (e.g. Pharmacia/LKB, BioRad, Gibco). The general procedure is as follows:

1. Cap the lower end of the appropriate number of gel tubes (a suitable size is 75 mm × 5 mm i.d.) and stand vertically.

2. Insert the gel mixture (solution (f)) into each tube by means of a pasteur pipette until the meniscus reaches a point 0.5 cm from the top of the tube. Ensure that no small air bubbles are trapped in the tube.

3. Apply a layer of distilled water (about 3 mm high) on top of the acrylamide layer to prevent the formation of a curved meniscus at the origin, taking care not to disturb the boundary; this is most easily achieved by allowing water to run slowly down the side of the tubes from a pasteur pipette.

4. Leave the filled tubes in a undisturbed position at room temperature for polymerization to occur (about 40 min). It will be observed that the initial interface between the acrylamide mixture and distilled water disappears a few minutes after pouring as a result of diffusion. This is later replaced by another interface at a position about 1 cm below the original boundary and represents the top of the polymerized gel.

5. Remove the water layer by rapidly flicking the inverted tube and take off the cap from the lower end.

6. Place the tubes in the electrophoresis tank.

7. Dilute solution (d) (reservoir buffer stock) ten times with distilled water and place in the electrophoresis tank. Ensure that no air bubbles are present in the top of the gel tubes.

8. With an automatic pipette, gently layer 20 μl of the apolipoprotein sample solution on top of the acrylamide layer in each gel tube.

9. Apply a current of 2 mA per tube until the tracking dye has reached the bottom of the tube (approximately 90 min).

10. Turn off the current and take the tubes out of the tank. Remove the acrylamide rods from the glass tubes by the application of gentle pressure to one end of the tube from a rubber teat after squirting water between the gel surface and the tube wall with a long, fine, blunt needle (e.g. 2.5 inch, 23 gauge) attached to a syringe. If R_F measurements are to be made, a short strip of metal wire should be inserted into the gel at the position of the tracking dye as this disappears on washing the gel.

11. Stain the gels by the method described for gradient gels in Section 4.1.3.

5.2.4 Evaluation of electrophoresis results

Identification of the apolipoprotein bands obtained on SDS gels may be made by comparing their molecular weights with those of the standard proteins. Plot the log of the mol. wt. of each marker against the R_F value (distance migrated by the proteins divided by distance moved by the tracking dye) for each of the marker proteins and read off the values for the apolipoproteins from this line (*Figure 5*). Bands separated by electrophoresis in 8 M urea are identified by reference to published patterns (see *Figure 5*) or by comparison of their R_F values with those of purified apolipoproteins. Identification of the minor apolipoprotein constituents of HDL subfractions by these procedures is by no means an easy task, largely because of the close proximity of some components to each other. The assignments made should therefore be confirmed either by the use of immunological methods (see Section 5.3 below) or by determination of the partial amino acid sequence of the apolipoproteins removed from excised regions of the gel containing them by electroelution and comparing their sequences with published sequence data (47).

Useful information on the composition and distribution of apolipoproteins may be obtained by visual observation of the stained electrophoresis gels. Semi-quantitative data is given by measurement of peak areas after scanning (see Section 4.1.4), but it must be emphasized that each apolipoprotein has its own distinct absorption characteristics, so that the proportion of peak areas measured does not represent the actual distribution of concentrations of the component apolipoproteins. If possible, calibration curves should be determined for each apolipoprotein.

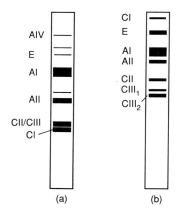

Figure 5. Separation of apolipoproteins from delipidized HDL on a (a) SDS–10% polyacrylamide gel and (b) 8 M urea–10% polyacrylamide gel.

5.3 Determination of apolipoprotein composition by immunological methods

Many of the problems associated with the use of electrophoresis for the quantitative determination of the apolipoprotein composition of HDL and its subfractions may be circumvented by the application of specific immuno-assays. These assays may be applied to the intact HDL particle and they therefore avoid the technical and theoretical difficulties that arise through delipidization and are inherent in all staining procedures. These methods include ELISA (enzyme-linked immunosorbent assay), immuno-nephelometry, radial immunodiffusion, and radioimmunoassay and are described in detail in Section 5 of Chapter 3.

It should be stressed here again that there is a considerable degree of inter-method variation in the determination of apolipoproteins so that no method, whether it be an immunoassay or an electrophoretic procedure, is yet capable of providing absolute apolipoprotein compositions. Specific immunoreaction also provides the ideal way to identify apolipoprotein bands on acrylamide gels after these have been transferred to a suitable membrane (see Chapter 3, Section 4).

5.4 Lipid analysis and particle composition

To determine the composition of HDL subfractions that have been separated by the above procedures, dialyse the appropriate fractions against 0.9% NaCl and remove aliquots from a single pool of each subfraction for the following determinations.

5.4.1 Total cholesterol and free cholesterol

The concentration of these components is most conveniently determined by enzymic assay using commercially-available kits (Boehringer, Merck), and following the manufacturer's instructions. Cholesteryl ester is calculated as the difference between total cholesterol and free cholesterol. The principle of the procedure and some practical considerations are discussed in Chapter 1.

5.4.2 Triglyceride

Enzyme assay kits for triglyceride determination are also available from the above manufacturers (see Chapter 1).

5.4.3 Phospholipid

Several chemical and enzymatic methods are available for the determination of the phospholipid content of lipoproteins. The underlying principle of most of the methods is the oxidation of phospholipids to phosphate with perchloric acid and hydrogen peroxide and the colorimetric assay of phosphate by the formation of a complex with molybdate (and vanadate) in the presence of nitric acid. An alternative procedure involves phospholipolysis and the release of choline. This is oxidized to betaine with the production of hydrogen peroxide and coupling of the latter to the production of a quinone chromogen. Test kits based on both of the procedures are produced by Boehringer.

A reliable lipid control standard should be included with each of the above determinations (e.g. Precinorm Lipid Control Serum, Boehringer).

5.4.4 Total apolipoprotein protein

A popular method is that based on Miller's modification (48) of the Lowry procedure (49).

6. Low density lipoproteins: Introduction

LDL contains a large proportion of the cholesterol present in the plasma of adult humans, and elevated concentrations are considered to be a major factor in the development of premature atherosclerosis. LDL is derived from VLDL which is synthesized in the liver and is converted by the action of lipoprotein lipase, situated on the blood capillary walls of adipose tissue, muscle, and other organs, to form intermediate density lipoproteins. A large part of these, about half in the case of humans, is normally processed further to form LDL. The conversion of VLDL into LDL results in the loss of triglyceride, phospholipids, and apolipoproteins other than apo B100 which remains as the major apolipoprotein associated with the resulting LDL particles. LDL is cleared from the plasma by a regulated pathway involving endocytosis following binding to apo B,E receptors present in the liver and other tissues. *In vitro* studies have shown that chemically modified LDL and LDL which has been biologically modified by incubation in the presence of endothelial cells are taken up through a system that is insensitive to control by modified LDL receptors on macrophages. This biological modification is believed to involve the oxidative breakdown of LDL phospholipids and apo B100; since foam cells found in early atherogenic lesions are derived largely from monocyte/macrophages, this process provides an intriguing hypothesis for the underlying cause of atherogenesis.

As is the case with HDL, LDL has now been clearly demonstrated to consist of discrete subspecies. Several laboratories have succeeded in separating LDL into a number of subfractions across its density range of 1.019 to 1.063 g/ml and it has been shown that the resolved fractions differ both in particle size and composition and in their metabolic activity (see Chapter 5). The observation that the distribution of LDL subfractions differs among individuals and between normal and hyperlipidaemic subjects, together with the suggestion that a specific subfraction may provide a risk marker for coronary heart disease, would indicate that this is likely to be an active area of future research (50–54).

7. Techniques for the resolution of LDL subfractions

Methods for the separation of LDL into a limited number of subfractions using density gradient ultracentrifugation techniques have been reported from relatively few laboratories (see reference 55 for a general practical account of the method). The procedures differ appreciably both in the construction of the gradient and in the fractionation profile obtained and no single version of the method has yet assumed general useage. Outlines of several methods that produce satisfactory separations will therefore be described; the method adopted may be determined by the range of ultracentrifuges and swing-out rotors that are available to the investigator. A good quality density former and means of collecting gradient fractions is highly desirable for a reliable and reproducible fractionation to be achieved.

Characterization of the isolated LDL subfractions is made by means of gradient gel electrophoresis which, on its own, lacks the resolving power for the quantitative assessment of LDL subfraction concentrations. The combined use of the two techniques is therefore advisable for a full evaluation of the distribution of LDL subfractions to be appreciated.

Methods for the separation of LDL subfractions of animal and avian species are also available (56, 57).

7.1 Density gradient ultracentrifugation

7.1.1 Method of Krauss and Burke (51)

This was among the first density gradient procedures to be described for the separation of LDL subfractions and in many respects forms the basis of later methods. This fractionation technique yields up to four density classes, designated LDL_1 to LDL_4 and is described in *Protocol 13*.

Protocol 13. Separation of LDL subfractions by density gradient ultracentrifugation by the method of Krauss and Burke (51)

1. Isolate LDL from plasma by ultracentrifugation between densities 1.009 and 1.063 g/ml using a modification of the procedure used in *Protocol 1*.

2. Dialyse overnight against two changes of NaBr solution of density 1.0400 g/ml.

3. Carefully layer 2 ml of the dialysed LDL above a NaBr solution of density 1.0540 g/ml (2.5 ml) in a 9/16″ × 3″ Beckman polyallomer or ultraclear centrifuge tube and layer 2.5 ml of NaBr solution of 1.0275 g/ml above the LDL.

4. Centrifuge the tubes at 40 000 r.p.m. for 40 h in a Beckman SW 41 rotor[a] at 22–24 °C.

Protocol 13. *Continued*

5. Withdraw the contents of the tubes, beginning with the top 0.5 ml, then 6 1-ml fractions and the bottom 0.5 ml.

^a These conditions are equivalent to those used in the original method which employed the now-obsolete SW45 rotor.

7.1.2 Method of Chapman *et al.* (52)

This procedure employs a modified and extended density gradient which separates LDL into eight or more fractions which vary in electrophoretic mobility, flotation rate in the analytical ultracentrifuge, particle size, and lipid and apolipoprotein composition (see *Protocol 14*).

Protocol 14. Density gradient separation of LDL subfractions by the method of Chapman *et al.* (52)

1. Separate LDL from plasma by ultracentrifugation between densities 1.099 and 1.063 g/ml and dialyse against NaBr solution of density 1.040 g/ml.

2. Using an Auto-Densiflow II (Buchler Instruments), or equivalent density gradient former, coupled to a peristaltic pump, pump 4.5 ml of a NaCl–KBr solution of density 1.054 g/ml into the bottom of a 9/16″ × 3″ Beckman Ultraclear centrifuge tube (capacity 13.2 ml). Then layer the following solutions onto the latter:

   • 3.5 ml of the dialysed LDL solution containing up to 15 mg protein

   • 2.0 ml of NaCl–NaBr solution of density 1.024 g/ml

   • 2.0 ml of NaCl solution of density 1.019 g/ml

3. Centrifuge in the Beckman SW 41 rotor at 40 000 r.p.m. for 44 h at 15 °C.

4. Fractionate the gradient with a density gradient fractionator (Model 185; ISCO) by puncture of the bottom of the tube and upward displacement of the gradient using a non-miscible, dense solution (e.g. Fluorinert FC 40; ISCO). Collect one fraction of 0.84 ml, followed by 15 fractions of 0.8 ml with a fraction collector.

7.1.3 Method of Griffin *et al.* (54)

This recently-reported density gradient procedure permits the separation of LDL subfractions directly from plasma within 24 h and is probably the method of choice at the present time. It divides LDL into three subfractions, LDL_1, LDL_2, and LDL_3, the concentrations of which differ in males and females and in patients with coronary heart disease. An advantage of this system is that it fractionates fresh plasma, since LDL subfractions tend to be unstable during prolonged centrifugation. The method also resolves subfractions that correspond to those defined by Krauss and Burke (51). The procedure is described in *Protocol 15*.

Protocol 15. Density gradient separation of LDL subfractions by the method of Griffin *et al.* (54)

1. Collect blood samples into EDTA (1 mg/ml) and adjust the density of the separated plasma to 1.09 g/ml by the addition of solid KBr (0.38 g/3 ml plasma).

2. Introduce the following solutions sequentially into polyvinyl alcohol-coated polyallomer SW-40 (9/16″ × 3 3/4″) Beckman tubes by peristaltic pump.

   - 0.5 ml of KBr solution of density 1.182 g/ml
   - 3.0 ml of the dialysed LDL solution (density 1.09 g/ml)
   - 1.0 ml of KBr solution of density 1.060 g/ml
   - 1.0 ml of KBr solution of density 1.056 g/ml
   - 1.0 ml of KBr solution of density 1.045 g/ml
   - 2.0 ml of KBr solution of density 1.034 g/ml
   - 2.0 ml of KBr solution of density 1.024 g/ml
   - 1.0 ml of KBr solution of density 1.019 g/ml

3. Accelerate the rotor to 170 r.p.m. over 4 min in a Beckman 48–60 ultracentrifuge and then centrifuge at 40 000 r.p.m. for 24 h. Stop the rotor with the brake off.

4. Displace the gradient containing the separated LDL fractions by upward displacement from the tube by infusion of a dense, hydrophobic material (e.g. Maxidens, 1.9 g/ml, Nycomed Ltd) under the plasma layer at a flow rate of 0.69 ml/min. Continuously monitor the eluate at A_{280}.

5. Measure the gradient in tubes in which the plasma was replaced by a NaBr solution of density 1.09 g/ml.

6. Identify the major LDL subfractions by peak maxima that occur between hydrated density intervals of 1.025–1.034 g/ml (LDL_1), 1.034–1.044 g/ml (LDL_2), and 1.044–1.060 g/ml (LDL_3). Minor subfractions may be distinguished as distinct peaks or shoulders in the absorbance profile.

7. Determine the concentrations of the individual LDL subfractions by proportioning the mass concentration of LDL between the peak areas and converting absorption units to lipoprotein mass equivalent by applying specific extinction coefficients (LDL_1, 1 optical density unit = 2.63 mg lipoprotein/ml; LDL_2, 1 OD unit = 2.94 mg/ml; LDL_3, 1 OD unit = 1.92 mg/ml).

7.1.4 Single spin gradient ultracentrifugation method of Swinkels *et al.* (58)

In this method, two LDL subfractions are clearly distinguished by the banding pattern in the density gradient: a lighter LDL_1 subfraction, occasionally showing a subdivision into two bands, and a heavier LDL_2 band. The resolved subfractions differ in size and in chemical composition. This provides a simple and quick method for the identification and isolation of LDL subfractions from small amounts of serum.

7.2 Analysis of LDL subfractions by gradient gel electrophoresis

Samples of LDL or its subfractions are submitted to electrophoresis, stained, and scanned by the same procedure as that described for the analysis of HDL subfractions in Section 4.1.4, except that 2–16% acrylamide gradient gels (Pharmacia) are used in place of the 4–30% gels.

Acknowledgements

I gratefully acknowledge the valuable contributions made in the development of methods for the separation and analysis of lipoprotein fractions by many former and present colleagues, especially Bruce A. Griffin, John A. Rooke, Kate Watt, and Heather M. Wilson. I sincerely thank Mrs Karen Slesser for her skill and patience in typing this manuscript.

References

1. Eisenberg, S. (1984). *J. Lipid Res.*, **25**, 1017.
2. Reichl, D. and Miller, N. E. (1989). *Arteriosclerosis*, **9**, 785.
3. Ghosh, D. K. and Menon, K. M. J. (1987). *Biochem. J.*, **244**, 471.
4. Darbon, J. M., Tournier, J. F., Tauber, J. P., and Bayard, F. (1986). *J. Biol. Chem.*, **261**, 8002.
5. Hinsbergh, V. W. M., Scheffer, M., Havekes, L., and Kempen, H. J. M. (1986). *Biochim. Biophys. Acta*, **878**, 49.
6. Patsch, J. R., Karlin, J. B., Scott, L. W., Smith, L. C., and Gotto, A. M. (1983). *Proc. Natl. Acad. Sci. USA*, **80**, 1449.
7. Gordon, T., Castelli, W. P., Hjortland, M. C., Kannel, W. B., and Dawber, T. R. (1977). *Am. J. Med.*, **62**, 707.
8. Gofman, J. W., de Lalla, O., Glazier, F., Freeman, N. K., Lindgren, F. T., Nichols, A. V., Strisower, E. H., and Tamplin, A. R. (1954). *Plasma*, **2**, 413.
9. Mahley, R. W., Weisgraber, K. H., Innerarity, T., Brewer, H. B., and Assmann, G. (1975). *Biochemistry*, **14**, 2817.
10. Marcel, Y. L., Weech, P. K., Nguyen, T. D., Milne, R. W., and McConathy, W. J. (1984). *Eur. J. Biochem.*, **143**, 467.

11. Blanche, P. J., Gong, E. L., Forte, T. M., and Nichols, A. V. (1981). *Biochim. Biophys. Acta*, **665**, 408.
12. Mahley, R. W. and Holcombe, K. S. (1977). *J. Lipid Res.*, **18**, 314.
13. Hay, C., Rooke, J. A., and Skinner, E. R. (1978). *FEBS Letters*, **91**, 30.
14. Lagrost, L., Gambert, P., Boquillon, M., and Lallemant, C. (1989). *J. Lipid Res.*, **30**, 1525.
15. Rooke, J. A. and Skinner, E. R. (1979). *Internat. J. Biochem.*, **10**, 329.
16. Gidez, L. I., Miller, G. J., Burstein, M., Slagle, S., and Eder, H. A. (1982). *J. Lipid Res.*, **23**, 1206.
17. Hames, B. D. and Rickwood, D. (ed.) (1981). *Gel Electrophoresis of Proteins: A Practical Approach*. IRL Press Ltd, Oxford.
18. Lefevre, M., Goudey-Lefevre, J. C., and Roheim, P. S. (1987). *J. Lipid Res.*, **28**, 1495.
19. Laboratory Techniques (1983). Gradient gel electrophoresis. In *Polyacrylamide Gel Electrophoresis*. p. 7. Pharmacia Fine Chemicals, Uppsala, Sweden.
20. Anderson, D. W., Nichols, A. V., Forte, T. M., and Lindgren, F. T. (1977). *Biochim. Biophys. Acta*, **493**, 55.
21. Anderson, D. W., Nichols, A. V., Pan, S. S., and Lindgren, F. T. (1978). *Atherosclerosis*, **29**, 161.
22. Griffin, B. A., Skinner, E. R., and Maughan, R. J. (1988). *Metabolism*, **37**, 535.
23. Skinner, E. R., Watt, C., Reid, I. C., Besson, J. A. O., and Ashcroft, G. W. (1989). *Clin. Chim. Acta*, **184**, 147.
24. Griffin, B. A., Skinner, E. R., and Maughan, R. J. (1988). *Atherosclerosis*, **70**, 165.
25. Weisgraber, K. H. and Mahley, R. W. (1980). *J. Lipid Res.*, **21**, 316.
26. Wilson, H. M., Griffin, B. A., Watt, C., and Skinner, E. R. (1992). *Biochem. J.* (in press).
27. March, S. C., Parikh, I., and Cuatrecasas, P. (1974). *Anal. Biochem.*, **60**, 149.
28. Klör, H. U., Schmidt, J. W., Ditschuneit, H., and Glomset, J. (1976). In *Protides of the Biological Fluids* (ed. H. Peters), Vol. 23, p. 581. Pergamon Press, Oxford.
29. Harris, E. L. V. (1989). In *Protein Purification Methods: A Practical Approach* (ed. E. L. V. Harris and S. Angal). IRL Press, Oxford.
30. Cheung, M. C. (1986). In *Methods in Enzymology* (ed. J. J. Albers and J. P. Segrest), Vol. 129, p.1 30. Academic Press Inc., London.
31. Gibson, G. C., Rubinstein, A., Ginsberg, H. N., and Brown, V. G. (1986). In *Methods in Enzymology* (ed. J. J. Albers and J. P. Segrest), Vol. 129, p. 186. Academic Press Inc., London.
32. Cheung, M. C. and Wolf, A. C. (1989). *J. Lipid Res.*, **30**, 499.
33. Atmeh, R. F., Shepherd, J., and Packard, C. J. (1983). *Biochim. Biophys. Acta*, **751**, 175.
34. James, R. W., Proudfoot, A., and Pomelta, D. (1989). *Biochim. Biophys. Acta*, **1002**, 292.
35. Clifton, P. M., MacKinnon, A. M., and Barter, P. J. (1987). *J. Chromatog.*, **414**, 25.
36. Kostner, G. M. (1978). In *Protides of the Biological Fluids* (ed. H. Peeters), Vol. 25, p. 83. Pergamon Press, Oxford.
37. Weisgraber, K. H. and Mahley, R. W. (1986). In *Methods in Enzymology* (ed. J. J. Albers and J. P. Segrest), Vol. 129, p. 145. Academic Press Inc., London.

38. Forte, T. M. (1986). In *Methods in Enzymology* (ed. J. P. Segrest and J. J. Albers), Vol. 128, p. 442. Academic Press Inc., London.
39. Hamilton, R. L., Williams, M. C., Fielding, C. J., and Havel, R. J. (1976). *J. Clin. Invest.*, **58**, 667.
40. Reichl, D., Forte, T. M., Hong, J. L., Rudra, D. N., and Pflug, G. (1985). *J. Lipid Res.*, **26**, 1399.
41. Osborne, J. C. (1986). In *Methods in Enzymology* (ed. J. J. Albers and J. P. Segrest), Vol. 128, p. 213. Academic Press Inc., London.
42. Scanu, A. M. and Edelstein, C. (1971). *Anal. Biochem.*, **44**, 576.
43. Shore, V. and Shore, B. (1967). *Biochemistry*, **6**, 1962.
44. Weber, K., Pringle, J. R., and Osborne, M. (1972). In *Methods in Enzymology* (ed. C. H. W. Hirs and S. N. Timasheff), Vol. 26, p. 3. Academic Press Inc., London.
45. Stephens, R. E. (1975). *Anal. Biochem.*, **65**, 369.
46. Davis, B. J. (1964). *Ann. N.Y. Acad. Sci.*, **121**, 404.
47. Brewer, H. B., Ronan, R., Meng, M., and Bishop, C. (1986). In *Methods in Enzymology* (ed. J. P. Segrest and J. J. Albers), Vol. 128, p. 223. Academic Press Inc., London.
48. Miller, G. L. (1959). *Anal. Chem.*, **31**, 964.
49. Lowry, O. H., Rosebrough, N. J., Farr, A. L., and Randall, R. J. (1951). *J. Biol. Chem.*, **193**, 265.
50. Shen, M. M. S., Krauss, R. M., Lindgren, F. T., and Forte, T. M. (1981). *J. Lipid Res.*, **22**, 236.
51. Krauss, R. M. and Burke, D. J. (1982). *J. Lipid Res.*, **23**, 97.
52. Chapman, M. J., Laplaud, P. H., Luc, G., Forgez, P., Bruckest, E., Goulinet, S., and Lagrange, D. (1988). *J. Lipid Res.*, **29**, 442.
53. Luc, G., Chapman, M. J., DeGennes, J.-L., and Turpin, G. (1986). *Eur. J. Clin. Invest.*, **16**, 329.
54. Griffin, B. A., Caslake, M. J., Yib, B., Tait, G. W., Packard, C. J., and Shepherd, J. (1990). *Atherosclerosis*, **83**, 59.
55. Kelly, J. L. and Kruski, A. W. (1986). In *Methods in Enzymology* (ed. J. P. Segrest and J. J. Albers), Vol. 128 p. 170. Academic Press Inc., London.
56. Luc, G. and Chapman, M. J. (1988). *J. Lipid Res.*, **29**, 1251.
57. Hermier, D., Forgez, P., and Chapman, M. J. (1985). *Biochim. Biophys. Acta*, **836**, 105.
58. Swinkels, D. W., Hak-Lemmers, H. L. M., and Demacker, P. N. M. (1987). *J. Lipid Res.*, **28**, 1233.

5

Lipoprotein turnover and metabolism

ALLAN GAW, CHRISTOPHER J. PACKARD, and JAMES
SHEPHERD

1. Introduction

1.1 Lipoprotein metabolism

The study of lipoprotein kinetics has added significantly to our understanding of the regulation of the plasma lipid transport system and the aberrations that give rise to hyperlipidaemia. Lipoprotein metabolism can be divided into three inter-related pathways (*Figure 1*). The first, associated with the synthesis and catabolism of chylomicrons, is responsible for the absorption of dietary fat and delivery of triglyceride to peripheral tissues (such as adipose tissue and skeletal muscle) and cholesterol to the liver. The second involves the production of very low density lipoproteins (VLDL) from liver and their remodelling to intermediate and low density lipoproteins (IDL, LDL). By this pathway triglyceride and cholesterol are transported in the fasting state. The third is concerned primarily with the return of cholesterol from peripheral sites to the liver and is mediated by the high density lipoproteins (HDL). The kinetics of these processes have been followed using both lipid and apolipoprotein markers. The former were used in early experiments and provided information on total body turnover rates of cholesterol and triglyceride. The latter have yielded data on the behaviour of individual lipoprotein fractions. Chylomicrons, VLDL, IDL, and LDL contain apolipoprotein B as their major structural protein. This high molecular weight polypeptide exhibits an extended secondary structure that intercalates into the lipoprotein surface and remains there throughout the lifetime of the particle in the plasma. Hence it acts as a marker of a particle's metabolic fate, whereas all other lipoprotein constituents including the A, C, and E apolipoproteins exchange freely between particles, complicating interpretation of their metabolic properties.

A further complication to the study of the plasma lipoprotein transport system that has emerged in recent years is its structural heterogeneity.

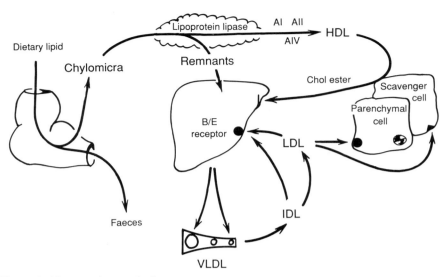

Figure 1. Lipoprotein metabolism.

Lipoproteins exist in plasma as a spectrum of particles of differing composition. They can be grouped into major classes but within each class high resolution techniques have shown the presence of subspecies. For example, LDL is readily subdivisible into LDL_1, LDL_2, and LDL_3, while HDL can be separated into distinct particle subtypes called HDL_2 and HDL_3. It is likely that these chemically distinct subfractions have differing metabolic properties whose understanding will hopefully contribute further to our knowledge of the role of lipoproteins in health and disease.

Apolipoproteins are synthesized primarily in the intestine and liver. In response to a dietary fat load, enterocytes will package lipid in the form of large triglyceride-rich chylomicron particles containing apo AI, AII, AIV, and the B48 isoform of apo B. Upon entry into the systemic circulation via the thoracic duct the A apolipoproteins are largely lost to HDL while apo C (CI, CII, CIII) and apo E are acquired by a reverse exchange process from HDL.

CII activates lipoprotein lipase to promote triglyceride hydrolysis which takes place in adipose and skeletal muscle capillary beds. During this process the core and surface of the chylomicron particle are reduced with further loss of apo A and apo C producing a remnant particle which is relatively rich in cholesterol esters and contains apo B48 and apo E on its surface. The liver, which recognizes the remnant particle by virtue of its apo E, assimilates it by an active transport mechanism.

The liver secretes triglyceride in the form of VLDL continuously throughout the day. The VLDL can vary in size from 70–35 nm (S_F 20–400) and this spectrum is conveniently subdivided into $VLDL_1$ (S_F 60–400) and

VLDL$_2$ (S$_F$ 20–60). These particles contain the larger isoform of apo B (B100), some apo C, and apo E. The latter are supplemented in the plasma by exchange from HDL.

Lipoprotein lipase is again responsible for the hydrolysis of large triglyceride rich VLDL into smaller particles, i.e. IDL. Apo B is conserved in this process. IDL are either removed directly from the plasma (probably via the B/E receptor route) or converted by the action of a second lipase (hepatic triglyceride lipase) to LDL. The latter is the major cholesterol transporting lipoprotein in the plasma and is responsible for the regulated delivery of sterol to those cells which require it for growth and hormone or bile acid production. The liver is the major site of LDL catabolism. When the receptor pathway is saturated LDL is removed by a putatively atherogenic mechanism which involves the activity of scavenger cells of the monocyte macrophage line.

HDL acts to capture cholesterol from extrahepatic cells including macrophages and delivers it either directly or indirectly (via exchange to LDL) to the liver.

1.2 Approaches to the study of lipoprotein kinetics

Early studies of plasma cholesterol and triglyceride kinetics were based on the use of labelled precursors ([^{3}H]glycerol, [^{14}C]fatty acids) that were incorporated into the plasma lipid fractions. Measurement of their clearance from the circulation gave preliminary crude information on whole body turnover rates. However the limitations of this approach became apparent when it was realized that lipids in the bloodstream were associated with specific proteins (as lipid–protein complexes or lipoproteins) which are capable of exchange between the different lipoprotein species. Consequently, most of the recent metabolic studies have used an exogenous labelling procedure in which the lipoprotein or apolipoprotein under study is isolated and trace-labelled usually by iodination. The tracer is then introduced back into the subject and frequent timed blood samples are drawn to follow its catabolism. The plasma disappearance rate of radioactivity from the protein associated with a specific lipoprotein fraction is used as an estimate of the fractional catabolic rate (FCR) of the protein; this is expressed in 'pools catabolized per day'. Determination of the plasma pool itself (in mg) permits calculation of the absolute catabolic rate of the protein (in mg per day) which is usually expressed per kilogram of body weight. In 'steady state' this is numerically equal to the lipoprotein apolipoprotein synthetic rate. This method was first applied to lipoproteins in the early investigations of Scott and co-workers (1) and Gitlin *et al.* (2) but most of the present day work is traced back to the studies by Langer, Strober, and Levy in the 1970s (3). They were first to isolate a defined lipoprotein class (LDL) and determine the synthetic and catabolic rates of its protein component (apolipoprotein B).

The validity of the exogenous labelling techniques is dependent on the following assumptions.

(a) The labelling process does not alter the structure of the lipoprotein under study, that is tracer and tracee are metabolically equivalent.

(b) The tracer is structurally and metabolically homogeneous.

(c) Catabolism occurs in or close to the plasma compartment or in a pool which is in rapid equilibrium with plasma.

(d) The concentration (pool) of the tracee plasma lipoprotein does not change during the period of study. That is, there is no long term trend or periodic rhythm.

Furthermore the methods used to analyse the protein radioactivity data can significantly affect the final interpretation of the results. A number of mathematical procedures have been used (4), including curve peeling, deconvolution analysis, interactive simulation, and multicompartmental modelling. The last, originally applied to lipoprotein systems by Berman, is the most versatile and can be used as a tool to test quantitative hypotheses about the system under study. The use of labelled amino acid precursors to follow apolipoprotein kinetics has not been a widely applied technique because of reluctance to give the large doses of [14C] needed to obtain meaningful results. Fisher (5) has provided some data on the incorporation of [14C]leucine into apo B containing lipoproteins while Eaton *et al.* (6) used [75Se]selenomethionine. Synthetic rates can be measured directly with labelled precursors if a good estimate of the immediate (usually hepatic) precursor amino acid pool can be obtained. Re-cycling of the tracer species usually limits the application of these methods to rapidly turning over species like VLDL. Complex computer modelling which accounts for re-utilization of labelled amino acids is required if LDL and HDL are studied. New developments in stable isotope detection systems have opened up the possibility that non-radioactive [13C]- or [15N]-enriched amino acids can be used for measuring synthesis rates. Early results with this technology are encouraging and provide data that closely mirror those obtained with radioiodinated tracers. It is of course still hampered by the recycling problem that is seen with radioactive endogenous precursors. The information it provides should be considered complementary to that acquired with the exogenously labelled tracers. It is difficult to see how stable isotope technology would ever completely replace the use of the latter.

The sections which follow detail protocols (mainly from our laboratory) for the study of lipoprotein apolipoprotein kinetics. Since the approach used for each lipoprotein class differs slightly they are presented individually. A further section then deals with methods of analysis.

2. Turnover protocols

A number of similarities exist between protocols but each has been presented separately in full to minimize confusion and assist the investigator in following the methods precisely.

2.1 LDL turnover

This technique, for assessing LDL kinetic parameters, is well established in a number of laboratories. The protocol is currently in use in our laboratory and is basically the same as that initially developed by Langer *et al.* (3). An adaptation of the technique has allowed the discrimination of receptor-mediated and receptor-independent catabolism of LDL. This modification is described later (see Section 2.1.3).

Protocol 1. LDL isolation

1. Obtain 50 ml of plasma from the subject after a 14 h fast.

2. Prepare LDL (*d* 1.030–1.050 g/ml) by rate zonal ultracentrifugation according to the method of Patsch *et al.* (7) or by density gradient centrifugation at the limiting densities quoted above.

3. Remove salt by dialysing against 0.15 M NaCl/0.01% Na_2EDTA (ethylenediaminetetra acetate), (pH 7.0) and if necessary concentrate the lipoprotein fraction by pressure filtration through an XM 100 A cellulose membrane (Amicon) to a protein concentration of 3–5 mg/ml (as determined by the procedure of Lowry *et al.* (8)).

Radiolabelling is carried out by the iodine monochloride method of McFarlane (9) as modified by Shepherd *et al.* (10).

Protocol 2. Labelling of LDL

1. Prepare the following iodination mixture: Mix 1.0 ml of LDL solution with 0.5 ml of 1.0 M glycine, pH 10.00, and 2.0 mCi of carrier-free $Na^{125}I$ (or $Na^{131}I$).

2. Add an appropriate volume of iodine monochloride solution (25 mM in 1.0 M NaCl) to yield an ICl:protein ratio of approximately 20 mol: 500 000 Da (Daltons) of protein. Mix gently (this introduces an iodine atom into tyrosine residues at a level of less than 10 mol iodine/molecule of LDL protein).

3. Separate free and bound radioiodide by passing the iodination mixture over a 1.0 × 10 cm column of Sephadex G-25 (PD10 column, Pharmacia). Elute with 0.15 M NaCl containing 0.01% Na_2EDTA, pH 7.0.

2.1.1 Sterilization and measurement of radioactivity

Sterilize the labelled tracer by membrane filtration through a 0.22 μm filter (Acrodisc, Gelman Sciences) which has been primed with the subject's unlabelled LDL. This should be done immediately prior to re-injection into the subject.

The radioactivity concentration (μCi/ml) should be determined after sterilization by counting a 10 μl aliquot of labelled LDL and comparing with ^{125}I or ^{131}I simulated standards.

2.1.2 Subject preparation

In all studies it is essential to ensure that thyroidal uptake of radioiodide has been blocked by the oral administration of potassium iodide (60 mg thrice daily). This regimen should be commenced three days prior to injection of radiolabelled lipoproteins and continued for the next 28 days.

Turnovers are conducted on an outpatient basis and the subjects are instructed to adhere to their regular diet and lifestyle. This ensures steady state conditions for the investigation of lipoprotein metabolism. Informed consent should be obtained from all subjects.

In the United Kingdom DHSS authorization is required for all radioactive tracer studies undertaken in man.

If the initial collection of fasting blood is done at 08.00 h it is possible using the rate zonal centrifugation technique to have the labelled LDL ready for re-injection within the same working day.

Protocol 3. Injection and sampling procedures

1. Inject autologous labelled LDL (25 μCi radioactivity, 1.0–2.0 mg LDL protein) into a peripheral vein.

2. After a time lapse of 10 min collect the first blood sample from a peripheral vein in the opposite arm of the subject. Thereafter collect a 10 ml fasting venous blood specimen each morning for the next 14 days. Collect specimens in tubes containing K_2EDTA as anticoagulant to give a final concentration of 1 mg/ml.

3. Collect continuous 24 h urine collections timed to 08.00 h over the 14 day turnover study period to estimate urinary excretion rates and urinary: plasma radioactivity ratios.

Protocol 4. Calculation of pool sizes

1. Prepare LDL from serial 4.0 ml fasting plasma samples obtained on days 3, 7, 10, and 13 of the turnover study by ultracentrifugation at limit densities *d* 1.010–1.063 g/ml.

2. Measure total protein and tetramethylurea (TMU) soluble protein (11) in each LDL sample after dialysis against normal saline.

3. Calculate apo B concentration as the difference between total protein and TMU-soluble protein and express as mg/100 ml. Correct for losses by relating recovered radioactivity to that seen in plasma.

4. Derive pool size for LDL apo B from the product of plasma volume (obtained by isotope dilution) and mean LDL-apo B plasma concentration.

   An alternative apo B pool size assay has been attempted by measuring the composition of LDL (d = 1.030–1.050 g/ml) and relating the cholesterol/protein content to the LDL cholesterol value estimated by the LRC protocol (12).

Protocol 5. Data collection and handling

1. Obtain plasma from each fasting blood sample by low speed centrifugation (4 °C).

2. Count ^{131}I and ^{125}I in 2 ml aliquots of each sample in an automated gamma-counter for 10 min.

3. Count ^{131}I and ^{125}I in 2 ml aliquots of each (well mixed) 24 h urine collection. Determine urine volumes and hence total urine radioactivities.

4. Construct plasma radioactivity decay curves and calculate urine:plasma ratios for each time point.

2.1.3 Modification of the protocol to measure receptor-mediated and receptor-independent LDL catabolism

A second tracer of chemically modified LDL is used to estimate LDL clearance by receptor-independent pathways (13). Chemical modification by cyclohexanedione blocks the arginine residues on LDL apo B which are important in receptor binding. Uptake of this tracer is therefore abolished.

Protocol 6. Measurement of receptor-mediated and receptor-independent pathways

1. Prepare LDL and label aliquots with ^{131}I and ^{125}I respectively as above (*Protocol 2*) with the exception that the eluting buffer for the PD-10 column should now be adjusted to pH 8.1.

2. Add 1.0 ml [^{131}I]LDL at a protein concentration of 3–5 mg/ml in a solution of 0.15 M NaCl/0.01% Na$_2$EDTA to 2.0 ml of freshly prepared 0.15 M 1,2-cyclohexanedione in 0.2 M sodium borate buffer, pH 8.1.

Protocol 6. *Continued*

3. Incubate this mixture for 2 h at 35 °C in a water bath.

4. Separate modified LDL from unbound reagents by passing the mixture over a 1.0 × 10 cm column of G-25 Sephadex (PD10 column, Pharmacia) and elute with 0.15 M NaCl/0.01% Na_2EDTA, pH 8.1.

5. Sterilize the [^{131}I]-modified LDL and [^{125}I]-native LDL separately, as above (Section 2.1.1).

6. Determine their radioactivity content and inject appropriate doses in rapid sequence through an indwelling catheter. Separate the injections with a bolus of sterile normal saline.

2.2 HDL turnover

The protocol which follows may be used to measure the kinetic parameters of apolipoprotein AI and AII metabolism and involves exogenous apolipo-protein labelling as described by Shepherd *et al.* (14). Apo AI or apo AII turnovers may be performed separately, or simultaneously using different labels as described below.

Protocol 7. Preparation of stock apo AI and apo AII

1. Collect 300 ml of fresh plasma from a proven HIV and HB_sAg-negative volunteer by plasmapheresis.

2. Add solid KBr (0.0834 g/ml) to increase the plasma density to 1.063 g/ml.

3. Centrifuge 35 ml aliquots of the preparation in a Beckman Ti60 ultracentrifuge rotor for 24 h at 40 000 r.p.m., 10 °C.

4. Recover VLDL + LDL by aspiration of supernatant from each tube and discard.

5. Harvest the infranatant HDL and raise its density to 1.21 g/ml by adding solid KBr (0.236 g/ml).

6. Centrifuge for a further 48 h at 40 000 r.p.m. and 15 °C in the above rotor.

7. Harvest the supernatant HDL by aspiration and dialyse exhaustively at 4 °C against 0.15 M NaCl/0.01 M Tris/0.01% Na_2EDTA, pH 7.6.

8. Lyophilize.

9. Delipidate the lyophilized HDL by using the following sequence of solvents.
   i. Ethanol/Diethyl ether   3:1 v/v.
   ii. Ethanol/Diethyl ether   3:2 v/v.
   iii. Diethyl ether
   Perform these steps at 4 °C over 72 h.

10. Dry the HDL apolipoprotein under N_2 and store at $-20\,°C$ prior to further processing.

11. Purify apo AI and apo AII from total HDL apolipoprotein by preparative HPLC.

12. Collect fractions corresponding to apo AI and apo AII (identified by 280 nm detector) and dialyse exhaustively against 0.05 M NH_4HCO_3.

13. Lyophilize and store at $-20\,°C$.

2.2.1 Labelling of apo AI and apo AII

Radiolabelling is carried out by the iodine monochloride method as modified by Bilheimer *et al.* (15).

To prepare the iodination mixture dissolve 1 mg of stock apo AI with 1 ml 0.1 M glycine buffer (pH 10), then add 1 mCi of carrier free $Na^{131}I$ and 28 µl of 5 nmol/µl iodine monochloride giving an ICl:protein ratio of 2.5:1. Mix gently.

Separate free from bound radioiodine by passing the iodination mixture over a 1.0×10 cm column of Sephadex G-25 (PD10 column, Pharmacia) and elute with Tris/Na_2EDTA buffer, pH 7.0. The column should be primed with the buffer prior to use.

For apo AII follow the above procedure but label with $Na^{125}I$ and use 18 µl of 5 nmol/µl ICl to yield the appropriate molar ICl:protein ratio.

Protocol 8. Reincorporation of labelled lipoproteins

1. Determine the radioactivity present in each labelled species (µCi/ml).

2. Add 250 µCi of $[^{125}I]AI$ and $[^{131}I]AII$ to 20 ml of plasma obtained freshly from the subject.

3. Incubate at room temperature for 1 h and mix continuously.

To isolate the labelled HDL, first adjust the density of the modified plasma to 1.21 g/ml by adding solid KBr (0.3 g/ml); then centrifuge in Beckman Ti60 rotor at 55 000 r.p.m. for 24 h at 15 °C.

2.2.2 Sterilization and determination of radioactivity

Sterilize the labelled tracers by membrane filtration through 0.22 µm disposable filters (Acrodisc, Gelman Sciences) which have been pre-primed with the subject's native plasma. This should be done immediately prior to re-injection into the subject. Radioactivity (µCi/ml) should be determined after sterilization by counting a 10 µl aliquot of labelled HDL and comparing with ^{125}I and ^{131}I simulated standards.

2.2.3 Subject preparation

See Section 2.1.2.

Protocol 9. Injection and sampling protocol

1. Inject labelled HDL into a peripheral vein.

2. After a time lapse of 10 min collect the first blood sample from a peripheral vein in the opposite arm of the subject. Thereafter collect a 10 ml fasting venous blood specimen each morning into tubes containing K_2EDTA as anticoagulant.

3. Collect continuous 24 h urine collections over the 14 day study period to facilitate the estimation of urinary:plasma radioactivity data.

Protocol 10. Data collection and handling

1. Obtain plasma from each blood sample by low speed centrifugation at 4 °C.

2. Count 2 ml aliquots of each plasma sample in an automated gamma-counter for 10 min.

3. Count 2 ml aliquots of each (well mixed) 24 h urine collection. Determine urine volumes and hence total urine radioactivities.

4. Construct radioactivity decay curves by plotting c.p.m. against time and calculate urine:plasma radioactivity ratios for each time point.

2.3 VLDL apo B turnover

The metabolism of large (S_F 60–400) and small (S_F 20–60) VLDL may be investigated following protocols published by Shepherd, Packard *et al.* (16, 17).

Protocol 11. Isolation of $VLDL_1$ and $VLDL_2$

1. Obtain 250 ml of plasma from the subject by plasmapheresis after a 14 h fast.

2. Prepare total VLDL ($d < 1.006$ g/ml) by ultracentrifugation in a Beckman Ti60 rotor for 24 h at 39 000 r.p.m. (10 °C).

3. Collect the supernatant VLDL and if the subject's plasma triglyceride level is greater than 2 mmol/litre dilute the VLDL solution with 0.15 M NaCl to a concentration corresponding to a plasma triglyceride level of approximately 1.5 mmol/litre. This is necessary to avoid VLDL$_1$ carry-over into the VLDL$_2$ subfraction.

4. Adjust the density of 12 ml of VLDL solution to 1.118 g/ml by the addition of solid NaCl (170 mg/ml).

5. Layer a 2 ml aliquot of this preparation over a 0.5 ml cushion of *d* 1.182 g/ml NaBr solution in six Beckman SW40 rotor tubes which have been precoated with polyvinylalcohol to reduce internal surface tension and permit better layering.

6. Construct a discontinuous salt gradient from *d* 1.0988–1.0582 g/ml above each according to *Figure 2*.

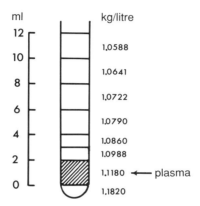

Figure 2. Discontinuous NaBr-gradient, used for the subfractionation of apo B containing lipoprotein by ultracentrifugation.

7. Centrifuge in a Beckman SW40 rotor at 39 000 r.p.m. for 1 h 38 min (23 °C) and decelerate without braking.

8. Remove VLDL$_1$ (S$_F$ 60–400) in the top 1.0 ml of solution and pool the material harvested from all tubes to yield 6 ml of VLDL$_1$ solution. Replace the volume removed with 1.0 ml of *d* 1.0588 g/ml solution before continuing with the separation.

9. Centrifuge at 18 500 r.p.m. (23 °C) for 15 h 41 min and decelerate without braking.

10. Remove VLDL$_2$ (S$_F$ 20–60) in the top 0.5 ml of solution for each of the six tubes and pool to yield 3 ml of VLDL$_2$ solution.

Radiolabelling is carried out by the iodine monochloride method as modified by Bilheimer *et al.* (15).

Protocol 12. Labelling of VLDL subfractions

1. Prepare the iodination mixture as follows:[a] To separate 2-ml aliquots of VLDL$_1$ and VLDL$_2$ solutions add 0.5 ml of 1.0 M glycine, pH 10.0 and 2 mCi of carrier-free Na^{131}I or Na^{125}I.

2. Add 6 µl of iodine monochloride solution (25 mM in 1.0 M NaCl) and mix gently.

3. Separate free from bound radioiodide by dialysing overnight against three times 2.0 litres of 0.15 M NaCl at 4 °C.

[a] These ratios of iodine/protein result in the incorporation of no more than one atom of iodine per molecule of apo B in VLDL.

2.3.1 Sterilization and measurement of radioactivity

Sterilize the labelled tracers by membrane filtration through 0.45 µm filters (Acrodisc, Gelman Sciences) which have been primed with the subject's native plasma. This should be done immediately prior to re-injection into the subject.

The radioactivity concentration (µCi/ml) should be calculated after sterilization by counting a 10 µl aliquot of labelled VLDL and comparing with ^{125}I and ^{131}I simulated standards.

2.3.2 Patient preparation

See Section 2.1.2.

On the third day after plasmapheresis the subject is admitted at 08.00 h following an overnight fast and the following procedure is observed.

Protocol 13. Injection and sampling protocol

1. Place an indwelling cannula in a peripheral vein to facilitate repeated venous blood sampling. Flush to maintain patency with 0.15 M NaCl— NOT with heparin containing solutions.

2. Inject autologous [^{131}I]VLDL$_1$ and [^{125}I]VLDL$_2$ in rapid sequence into a peripheral vein in the opposite arm. Separate injectates with a bolus of sterile normal saline.

3. Collect 10 ml venous blood samples via cannula at the following time points post-injection:[a] 10 min, 30 min, 1 h, 1.5 h, 2 h, 3 h, 4 h, 6 h, 8 h, 10 h, 14 h, and thereafter obtain 10 ml fasting venous samples each morning for the next 12 days. Collect all samples in tubes containing K$_2$EDTA as anticoagulant.

[a] The subject should remain fasting for the first 10 h of the study to minimize chylomicron production but is allowed unlimited non-caloric fluids.

Protocol 14. Reisolation of tracer

1. Obtain plasma from each fasting blood sample by low speed centrifugation (4 °C).

2. From 2 ml aliquots of plasma isolate $VLDL_1$ and $VLDL_2$ as described above (see *Protocol 11*, steps 4–10).

3. In addition isolate IDL (S_F 12–20) and LDL (S_F 0–12) by subjecting the loaded SW40 rotor to two further centrifugation steps of 39 000 r.p.m. for 2 h 35 min, and 30 000 r.p.m. for 21 h 10 min at 23 °C respectively. Remove IDL in the top 0.5 ml of the gradient and LDL in the top 1.0 ml.

4. Precipitate apo B by adding an equal volume of freshly redistilled 1,1,3,3-tetramethylurea at 37 °C to each lipoprotein fraction (11). Vortex and stand for 1 h.

5. Separate the resulting insoluble pellicle by centrifugation at 1000g for 30 min. Remove TMU-soluble phase carefully with a thinly drawn-out glass Pasteur pipette. Invert the tube to draw off any remaining liquid.

6. Delipidate the apo B pellet by extracting with organic solvents (ethanol: diethyl ether 3:1 v/v) overnight at −20 °C. Centrifuge to obtain a protein pellet. Remove the solvent.

7. Dry the pellet with a diethyl ether wash for 2 h at −20 °C and after removal of the ether place samples uncapped in incubator at 40 °C for 10 min or until ether has evaporated.

8. Redissolve the isolated apo B samples by adding 1.0 ml of 0.5M NaOH. Leave samples at 37 °C overnight or until dissolution is complete.

Protocol 15. Calculation of pool sizes

1. Prepare $VLDL_1$, $VLDL_2$, IDL, and LDL by the cumulative flotation ultracentrifugation of 6×2 ml aliquots of pooled plasma obtained on four occasions during the course of the 13 day VLDL-turnover study.

2. Measure total protein and TMU-soluble protein in each lipoprotein fraction.

3. Calculate apo B concentrations as the difference between total protein and TMU-soluble protein and express as mg/100 ml.

4. Derive pool sizes for apo B in the four lipoprotein fractions from the product of plasma volume and the plasma concentration of apo B in each fraction.

5. Correct for centrifugal losses by comparing the recovered $VLDL_1$ + $VLDL_2$ + IDL + LDL cholesterol to the 'non-HDL' cholesterol in plasma.

Protocol 16. Data collection and handling

1. Count ^{131}I and ^{125}I in each of the apo B samples prepared from the lipoprotein fractions.
2. Calculate specific activities by estimating the protein content of each sample by a modified Lowry method in which NaOH is omitted from the Cu/alkali reagent. The NaOH added to the apo B protein to facilitate its dissolution yields the appropriate pH in the final Lowry reaction mixture.
3. Use the 10 min sample to determine the initial level of radioactivity in each lipoprotein fraction and to determine the subject's plasma volume by isotope dilution. This should be equal to about 4% of the body weight (i.e. a 70 kg man has a plasma volume of 2800 ml).

2.3.3 Data analysis

(a) Multiply the apo B specific activities by the average pool size to generate total radioactivities. Express these as a fraction of the total apo B radioactivity present at 10 min after injection.

(b) Use these data to define apo B decay curves for each lipoprotein fraction.

(c) Using these curves and the apo B protein masses associated with each lipoprotein fraction, simulate apo B metabolism in a multicompartmental model using the SAAM 30 program (18). The model is depicted in *Figure 3* and the procedure is discussed further in Section 3.4.

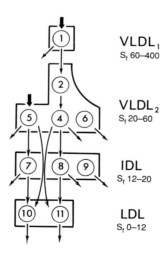

Figure 3. Model simulating apo B metabolism.

The model is a development of an earlier published version (17).

(a) Large VLDL apo B (VLDL$_1$, S$_F$ 60–400) behaves as a single species which decays monoexponentially in both normal and FH subjects. This is either catabolized directly or transferred to the VLDL$_2$ (S$_F$ 20–60) range.

(b) Within the S$_F$ 20–60 lipoproteins there is an arrangement akin to that described by Berman *et al.* (32). Some apo B enters a catabolic cascade and is converted to IDL (S$_F$ 12–20) while other material is diverted into a slowly metabolized remnant species (compartment 6).

(c) There is input of newly synthesized apo B into the S$_F$ 20–60 density range. This is required because not all of the S$_F$ 20–60 apo B mass can be accounted for by transport from large VLDL, and when large (S$_F$ 60–400) and small (S$_F$ 20–60) VLDL are labelled separately the kinetics of appearance of these tracers in IDL and LDL apo B is different. Usually the radioactivity derived from labelled small VLDL appears more quickly in these denser fractions and accounts for a higher proportion of their mass. Provision is made for this phenomenon by incorporating in the model parallel pathways leaving small VLDL and appearing in IDL and LDL.

(d) In the IDL range it is necessary to postulate the existence of a slowly metabolized species (compartment 9).

(e) LDL is distributed between two plasma compartments (compartments 10 and 11), only one of which is permitted to equilibrate with the extra vascular space. This is necessary to accommodate the observation that there are differential rates of appearance and removal of LDL apo B depending on whether the protein was derived from large or small VLDL.

2.4 Apo E turnover

In view of the polymorphism of apo E it is important to know the apo E phenotype of both the apolipoprotein donor and the subject under study. The protocol used in our laboratory is that of Gregg *et al.* (19) (see also Chapter 2).

Protocol 17. Isolation and labelling of apo E

1. Collect 250 ml of fresh plasma from a proven HIV and HB$_s$Ag-negative volunteer of known apo E phenotype by plasmapheresis.

2. Prepare VLDL ($d < 1.006$ g/ml) as described above.

3. Delipidate with chloroform/methanol (2:1 v/v).

4. Separate apo E from other apoliproteins by heparin affinity and Sephacryl S-200 gel permeation chromatography.

Protocol 17. *Continued*

5. Lyophilize apo E preparation.

6. Redissolve apo E in a buffer of 6 M guanidine–HCl and 1.0 M glycine, pH 8.5 to a concentration of 2 μg/μl.

7. To this mixture add 1–5 mCi $Na^{125}I$ or $Na^{131}I$.

8. Vortex mixture while slowly adding 3–15 μl of 0.33 mM iodine monochloride in 0.15 M NaCl.

Protocol 18. Reintroduction of labelled apo E

1. Incubate radiolabelled apo E with plasma (1–2 μl of iodination mixture/ ml plasma) for 30 min at 37 °C.

2. Adjust the plasma density to 1.21 g/ml by the addition of solid KBr.

3. Centrifuge in a Beckman Ti60 rotor at 55 000 r.p.m. for 24 h at 15 °C.

4. Isolate the $d < 1.21$ g/ml supernatant by tube slicing and dialyse against 0.15 M NaCl, 0.1 M Tris–HCl, pH 7.4, and 0.01% Na_2EDTA.

5. Add human serum albumin at a final concentration of 1%.

2.4.1 Sterilization and calculation of radioactivity

Sterilize the labelled tracer by membrane filtration through 0.2 μm filter (Acrodisc, Gelman Sciences) which has been pre-primed with the subject's native plasma. Radioactivity (μCi/ml) should be calculated after sterilization by counting a 10 μl aliquot of plasma and comparing with ^{125}I or ^{131}I simulated standards.

See Section 2.1.2 for subject preparation.

Protocol 19. Injection and sampling

1. Inject up to 25 μCi of ^{131}I radiolabelled or 100 μCi of ^{125}I radiolabelled apo E, approximately 5.5 h after the previous meal.

2. Collect 20 ml venous blood samples into K_2EDTA tubes at 10 min and at 6, 12, 18, 24, 36, and 48 h, and then daily until day 7. Collect all blood samples after a 6 h fast except the 10 min sample.

3. Collect continuous 24 h urine collections over the 7 day turnover period.

Protocol 20. Data collection and handling

1. Obtain plasma from each blood specimen by low speed centrifugation at 4 °C.

2. Add sodium azide and aprotinin to plasma at a final concentration of 0.5% and 200 KIU/ml, respectively, to prevent microbial growth and to inhibit serine protease cleavage of apo E.

3. Freeze 1.0 ml aliquots of plasma at −20 °C for each time point. Use these aliquots for apo E assay.

4. Prepare VLDL, IDL, LDL, and HDL fractions from each plasma sample by methods previously outlined.

5. Count radioactivity in equal aliquots of plasma, in the isolated lipoprotein fractions and in aliquots of the 24 h urine collections.

    Measure apo E mass at each time point in plasma and in lipoprotein subfractions by immunochemical techniques (see Chapter 3). Construct specific activity (c.p.m./μg apo E) decay curves.

2.5 Apo C turnover

The protocol described is basically that of Huff *et al.* (20).

Protocol 21. Isolation and labelling of VLDL apo C

1. Obtain fasting plasma from subject and isolate VLDL (S_F 20–400) by ultracentrifugation at *d* 1.006 g/ml.

2. Wash VLDL once by flotation through 0.15 M NaCl pH 7.4 containing 1.0 mM Na_2EDTA and 1.0 mM Tris buffer.

3. Radiolabelling is carried out by the iodine monochloride method as modified by Bilheimer *et al.* (15), see *Protocol 12*.

For sterilization and calculation of specific activity see Section 2.3.1.

Subject preparation is as in Section 2.1.2. Following this inject autologous labelled VLDL into a peripheral vein. After a time lapse of 10 min, collect the first blood sample from a peripheral vein in the opposite arm of the subject and thereafter collect samples throughout the next 48 h. All samples should be preserved in tubes containing K_2EDTA as anticoagulant.

To minimize chylomicron production the subject should receive a minimal fat diet throughout the 48 h period. This should comprise no more than 4% of calories as fat, 80% as carbohydrate, and 16% as protein.

Protocol 22. Reisolation of labelled apo C

1. Obtain plasma from each sample by low speed centrifugation 4 °C.

2. Isolate VLDL (*d* < 1.006 g/ml) and wash once by recentrifugation at the appropriate salt densities.

Protocol 22. *Continued*

3. Isolate total soluble VLDL apolipoprotein by the isopropanol precipitation method of Holmquist *et al.* (21).

4. Extract the soluble isopropanol–water phase with chloroform, methanol, and ether sequentially to precipitate soluble proteins and delipidate.

5. Centrifuge to pellet the apolipoprotein C.

6. Wash with diethyl ether.

7. Dry under N_2.

8. Redissolve pellet in 200 μl of buffer containing 8 M urea in 0.1 M Tris–HCl, pH 8.2.

Protocol 23. Separation of C apolipoproteins and determination of specific activity

1. Prepare polyacrylamide gels (7.5%) containing 6.8 M urea and 2% ampholine (LKB, pH 4–6).

2. Add 100 μl delipidated apo C solution per gel with 20 μl of a 1:5 dilution of ampholine (with 8 M urea) and 50 μl of 80% w/v sucrose.

3. Focus and stain according to the method of Swaney and Gidez (22).

4. Scan gels at 560 nm.

6. Quantitate apo C mass from standard curves prepared by isoelectric focusing of CII, CIII$_1$, and CIII$_2$ apolipoproteins (see Chapter 2).

6. After scanning, slice gel segments and count bands in an automated gamma-counter.

7. Calculate specific activities (c.p.m./μg protein) and construct decay curves for CII, CIII$_1$, and CIII$_2$.

2.6 Lipoprotein kinetics using stable isotope technology

The use of stable isotopes in biological research is not new. The first isotopic tracer studies conducted many years ago employed a stable isotope of hydrogen to study fat metabolism in mice. Their application to lipoprotein kinetic research is, however, more recent and has been facilitated by advances in mass spectrometry technology. The kinetic work performed so far has produced results which are comparable with those obtained in exogenous radiolabelling studies. The general problems inherent in endogenous labelling with radioactive precursors are also applicable to stable isotope technology but an important advantage in using stable isotopes is that of increased patient safety and acceptability. The analysis of stable isotopic data cannot, however, be viewed as directly analogous to that of radioisotope data (23).

Stable isotope precursors used to date in lipoprotein research include [^{15}N]glycine, [^{2}H]leucine, and [^{13}C]phenylalanine. If validated, the use of stable isotope protocols may extend the kinetic approach for the first time to the study of children and pregnant women and will make multiple studies in the same patient more acceptable.

Further refinement of protocols already published will undoubtedly take place in the light of recent efforts in this field. What follows may therefore be regarded as a general framework and is a modification of the work of Cryer *et al.* (24).

2.6.1 Subject preparation

The subject should be examined after a 14 h fast to minimize chylomicron synthesis and should remain fasted throughout the infusion period. Unlimited non-caloric fluids should however be encouraged to promote urine production.

Protocol 24. Injection and sampling

1. Immediately prior to injection obtain venous blood samples for the estimation of basal plasma precursor glycine pool and apolipoprotein enrichment with ^{15}N.

2. Administer a bolus injection of 0.15 mg/kg [^{15}N]glycine and immediately commence the continuous infusion of [^{15}N]glycine in 0.15 M NaCl at a rate of 0.15 mg/kg/h via a calibrated syringe pump.

3. Continue the infusion until steady state has been achieved (i.e. approximately 8–10 h).

4. Obtain further samples for lipoprotein isolation and measurement of glycine enrichment at 8, 9, and 10 h after the initial bolus injection.

   At each time point sufficient plasma should be obtained to store 2 × 5 ml aliquots for free plasma [^{15}N]-glycine estimation and to prepare 1–3 mg VLDL apo B.

5. Collect timed 2 h urine samples throughout the study period for the estimation of urinary [^{15}N]hippurate as a measure of precursor ^{15}N enrichment.

Protocol 25. Sample analysis

1. Prepare plasma from all samples by low speed centrifugation at 4 °C.

2. Freeze the 5 ml plasma aliquots at −20 °C for plasma precursor pool enrichment estimation (free plasma [^{15}N]glycine).

Protocol 25. *Continued*

3. Prepare VLDL-apo B by ultracentrifugation, TMU precipitation and delipidation as described above (see Section 2.3).

4. Store dried apo B pellets at −20 °C prior to further analysis (if necessary).

5. Hydrolyse apo B samples using 6 M HCl at 160 °C for 4 h. This is performed in an evacuated tube.

6. Isolate the glycine fraction from hydrolysates using preparative HPLC.

7. Lyophilize samples in aluminium boats prior to introduction into a suitable detection system. We have used a Continuous Flow Isotope Ratio-Mass Spectrometer (Europa Scientific Ltd).

---

3. Methods of analysis

All kinetic experiments in complex systems such as the human body present two problems. The first is how to collect the required data and the second is how to interpret it. The latter is often no less daunting than the former. Previous sections described detailed methods for gathering information on lipoprotein interconversion and catabolic rates; the discussion which follows outlines various mathematical analysis that can be applied to the raw data to extract physiologically meaningful parameters. All such analyses are approximations to the real situation. They involve assumptions that are often simple guesses as to what may be happening in the body.

3.1 Urinary excretion rate as a measure of catabolism

If the site of irreversible loss of trace-labelled material is restricted to the plasma or a compartment in rapid equilibrium with it then the ratio of urinary radioactivity to that in plasma is a measure of the fractional catabolic rate (FCR) of the lipoprotein. This urine/plasma ratio (U/P ratio) is valid if it can be demonstrated that:

- urine is the only site of loss
- the level of small-molecular weight degradation products in plasma is negligible
- the tracer is homogeneous
- urine collections are complete over each 24 h period

Radioiodinated plasma protein including LDL and HDL can usually be analysed by this method. Normally the total urine radioactivity in a 24 h period is measured and related to the mean plasma radioactivity during the same period. With radioiodinated tyrosine there is normally a 0.5 day delay between its appearance in the plasma and its excretion in urine. Thus a more correct measure of the mean plasma c.p.m. may be taken as the radioactivity at the *start* of the 24 h period. It is usual to collect urine continuously over the 14 days of an LDL or HDL turnover and examine the daily U/P ratio (*Figure 4*). Deviation from a steady value may be due to tracer damage (especially high peak values in the first two days) or more likely tracer heterogeneity. The latter is readily seen in most LDL turnovers in normal subjects but not in hypercholesterolaemics (25). The U/P ratio is a valuable parameter and can be viewed as a daily survey of the catabolic potential of the tracer. A mean of all 14 U/P values may be used as a measure of the FCR of LDL and it bears a close resemblance to the value obtained by analysis of plasma data. However, a recent suggestion (26) that single U/P ratio estimates at day 7 of an LDL turnover may be sufficient to determine an FCR precisely is not valid (27).

Urine data is a valuable source of information in LDL and HDL turnovers particularly if a multicompartmental approach is used in analysis as described below. It is an important marker of catabolism and should be included, where possible, in all protocols.

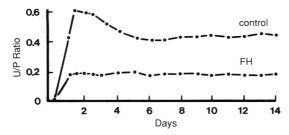

Figure 4 Urine/plasma radioactivity ratios determined over the 14 day period following injection of [^{125}I]LDL into a familial hypercholesterolaemic heterozygote or a control subject. The inconstancy of the ratio in the control subject probably reflects metabolic heterogeneity within his LDL spectrum. This phenomenon is less apparent in the FH heterozygote.

3.2 Determination of fractional catabolic rate from plasma data

When daily plasma radioactivities are plotted on semi-logarithmic graph paper, most lipoproteins and apolipoproteins display bi-exponential kinetics. That is the decay curve can be described by the equation $y = C_1 e^{-b_1 t} + C_2 e^{-b_2 t}$ where C_1 and C_2 are the intercepts on the y axis, and b_1 and b_2 are the slopes of the two exponentials. If the tracer is homogeneous and catabolized solely from the plasma compartment then a number of kinetic parameters can be determined by the mathematical approach of Matthews (28).

● the fractional catabolic rate

$$FCR = \frac{1}{C_1/b_1 + C_2/b_2}$$

- the percentage of intravascular tracer

$$\% \text{ intravascular} = \left[\frac{C_1}{b_1} + \frac{C_2}{b_2}\right]^2 \bigg/ \frac{C_1}{(b_1)^2} + \frac{C_2}{(b_2)^2}$$

- the capillary transfer rate

$$\text{CTR} = \frac{C_1 C_2 \, (b_2 - b_1)^2}{C_1 b_2 + C_2 b_1}$$

Practically, the plasma data are plotted to determine the start of the second, slow exponential. Simple regression can then be used to provide a line of best fit to this exponential, yielding b_1 and C_1. The first exponential is then extrapolated to the y axis and these projected values are subtracted from the plasma curve at suitable intervals (e.g. 0.5, 1.0, 2.0, 2.5, 3.0, 3.5, 4.0 days in an LDL turnover) to calculate the second exponential. Linear regression is used again to derive b_2 and C_2. If the second exponential is curvilinear then further 'curve-stripping' is needed and extra terms must be added to the kinetic equations but usually two exponentials are sufficient for analysis of LDL and HDL plasma decay curves.

Determination of the plasma fractional catabolic rate by either the Matthews approach or the U/P ratio permits an absolute catabolic rate to be calculated, i.e. FCR × mass of apolipoprotein in the circulation. If it can be demonstrated that the subject is in steady state then this is numerically equal to the synthetic rate of the apolipoprotein. Pragmatically we accept 'steady state' if

(a) the subject's weight has not altered over the period of study,

(b) there is no trend (i.e. gradient of > 10%) in the plasma concentration of the lipoprotein under study, and

(c) the coefficient of variation of serial lipoprotein estimations is less than 10%.

3.3 Analysis of precursor–product specific activity curves

Studies of lipoprotein interconversions such as the VLDL to LDL delipidation step have been analysed using a variety of techniques. A popular approach has been to calculate the specific activities (c.p.m./mg apo B) of the precursor lipoproteins (VLDL) and the products (LDL or IDL and LDL). The relationship between these curves reveals something of the metabolic links between the lipoprotein species.

Zilversmit (29) has shown that if a labelled precursor (A) is the sole source of the product (B) then the decay curve for the specific activity of A will cut the product curve at the maximum value (peak) of the latter. If the specific activity of B is less than the specific activity of A at the time of the peak then other 'cold' input is contributing to the mass of pool B. This simple analysis was used in early studies to establish the link between apolipoprotein B in VLDL and LDL. Furthermore, the areas under the precursor and product curves may be used to derive quantitative measures of the VLDL–LDL conversion rate (30).

A more sophisticated technique for mathematical analysis of specific activity data has been used by Sigurdsson *et al.* (31). Deconvolution or impulse analysis represents an extension of indicator dilution theory. It relies on

(a) obtaining disappearances curves for [^{125}I]VLDL-B and [^{131}I]LDL-B,

(b) measuring an appearance curve for [^{125}I]apo B in LDL, and

(c) unequivocal extrapolation of the precursor curves to infinity (i.e. to 'zero').

The above techniques which depend on the relationships between curves have been termed 'model independent' and to a certain extent this is valid since they make few assumptions about the system under investigation. The term, however, is misleading in that fundamental preconceptions such as precursor and product homogeneity are inherent in the functions used for analysis. The approach is also limited in the amount of information it can derive from a given data set.

3.4 Multicompartmental modelling

Data fitting using mechanistic models is a powerful, general approach to the study of lipoprotein kinetics. It has been used successfully in analysis of both lipid and apolipoprotein turnovers. Its main disadvantage is that detailed structured models have to be specified which are based on the physiological processes under investigation but often contain features necessary for the solution that cannot be attributed to a known metabolic step. The major tool that has been used in the area of lipoprotein metabolism is the Simulation Analysis And Modelling (SAAM) computer program generated by Berman and Weiss (18). It is a package that, given an observed data set and initial estimates, will alter the parameters in turn until it reaches a minimum sum of squares for the residual differences between observed and calculated data. A detailed explanation of the use of this complicated program is beyond the scope of the present discussion. However it is useful to provide some background on its potential application and its benefits over other approaches.

Multicompartmental modelling can be used for parameter estimation in complex metabolic pathways and is limited mainly by the quality and quantity of information contained in the data set. It also allows quantitative hypotheses to be generated and tested. We have developed a model to describe the metabolism of VLDL subfractions and their conversion to IDL and LDL. The information needed to generate parameter estimates includes: (a) Total apo B radioactivities for each fraction ($VLDL_1$, $VLDL_2$, IDL, and LDL). These are commonly expressed as the fraction of the total apo B radioactivity present in all four fractions at zero time. (b) Pool size measurements for each fraction. These are derived from the plasma volume (4% of body weight) and the serial determination of apo B in each individual fraction, corrected for centrifugal and handling losses. These data together with estimates of the initial conditions in each compartment and starting values for kinetic constants are incorporated into the SAAM problem deck. Manual fitting of calculated curves to experimental data is carried out until the solution is close to that required at which point the program is set to iterate automatically. Output from the SAAM program consists of parameter estimates and, importantly, their associated errors. The latter make possible an assessment of the accuracy of the rate constants and the validity of the model. This approach to kinetic analysis should not be reserved for large problems. Analysis of simple urine and plasma data is enhanced by the use of SAAM or similar programs. However the benefits of multicompartmental modelling should not be overstated. The final model used to define the system *may* be physiologically meaningful but is still a simplified mathematical description of a complex biological process.

4. Future prospects

Recent advances in our understanding of the genetics and substructure of the lipoprotein transport system has prompted a renewed interest in lipoprotein kinetics. The interindividual variation in response to diet or drugs appears to be linked to the recipient's genotype and hence some underlying metabolic variation in the population. We are challenged, therefore, to develop kinetic techniques which adequately describe the genetically based variability that is seen in both the normal and the hyperlipidaemic populations and to explain how subfractions of the major lipoprotein classes arise. These developments will require the integration of lipid and apolipoprotein turnovers and the ability to determine individual synthetic and catabolic rates for the various lipoprotein species. The availability of high resolution lipoprotein separation techniques such as gradient gel electrophoresis or density gradient ultracentrifugation will permit significant advances in this area. The recognition of new lipoprotein risk markers for coronary heart disease, e.g. Lp(a), and elucidation of their physiology and pathophysiology will depend heavily on an explanation of their metabolic behaviour in normal and diseased individuals.

Thus we see a continuing and indeed increasing need for an understanding of the metabolism of what is essentially a dynamic transport system.

Acknowledgements

We would like to acknowledge the secretarial expertise of Miss Patricia Price and Miss Claire McKerron who diligently prepared this manuscript. Portions of this work were performed under the tenure of grants BHF 187006 and BHF 817106.

References

1. Walton, K. W., Scott, P. J., Dykes, P. W., and Davies, J. W. L. (1965). *Clin. Sci.*, **29**, 217.
2. Gitlin, D., Cornwall, D. G., Nakasato, D., Oncley, J. L., Hughes, W. L., Jr, and Janeway, C. A. (1958). *J. Clin. Invest.*, **37**, 172.
3. Langer, T., Strober, W., and Levy, R. I. (1972). *J. Clin. Invest.*, **51**, 1528.
4. Berman, M., Grundy, S. M., and Howard, B. (ed.) (1982). *Lipoprotein Kinetics and Modelling.* Academic Press Inc., London.
5. Fisher, W. R., Zech, L. A., Bardalaye, P., Warmke, G., and Berman, M. (1980). *J. Lipid. Res.*, **21**, 760.
6. Eaton, R. P. and Kipnis, D. M. (1972). *Diabetes*, **21**, 744.
7. Patsch, J. R., Sailer, S., Kostner, G., Sandhofer, F., Holasek, A., and Braunsteiner, H. (1974). *J. Lipid Res.*, **15**, 356.
8. Lowry, O. H., Rosebrough, N. J., Farr, A. L., and Randall, R. J. (1951). *J. Biol. Chem.*, **193**, 256.
9. McFarlane, A. S. (1958). *Nature*, **182**, 53.
10. Shepherd, J., Bedford, D. K., and Morgan, H. G. (1975). *Clin. Chim. Acta*, **66**, 109.
11. Kane, J. P., Sata, T., Hamilton, R. L., and Havel, R. J. (1975). *J. Clin. Invest.*, **56**, 1622.
12. Lipid Research Clinics Program Manual of Laboratory Operations (1975) DHEW Publications, NIH No. 75–628, Government Printing Office, Washington DC.
13. Slater, H. R., McKinney, L., Packard, C. J., and Shepherd, J., (1984). *Arteriosclerosis*, **4**, 604.
14. Shepherd, J., Packard, C. J., Gotto, A. M., Jr, and Taunton, O. D. (1978). *J. Lipid. Res.*, **19**, 656.
15. Bilheimer, D. W., Eisenberg, S., and Levy, R. I. (1972). *Biochim. Biophys. Acta.*, **260**, 212.
16. Shepherd, J., Packard, C. J., Stewart, J. M., Atmeh, R. F., Clark, R. S., Boag, D. E., Carr, K., Lorimer, A. R., Ballantyne, D., Morgan, H. G., and Lawrie, T. D. V. (1984). *J. Clin. Invest.*, **74**, 2164.
17. Packard, C. J., Munro, A., Lorimer, A. R., Gotto, A. M., and Shepherd, J. (1984). *J. Clin. Invest.*, **74**, 2178.
18. Berman, M. and Weiss, M. (1978). 'SAAM Manual', DHEW Publ. NIH No. 78–180, US Govt Printing Office, Washington, DC.
19. Gregg, R. E., Zech, L. A., Schaefer, E. J., Start, D., Wilson, D., and Brewer, H. B., Jr (1986). *J. Clin. Invest.*, **78**, 815.

20. Huff, M. W., Fidge, N. H., Nestel, P. J., Billington, T., and Watson, B. (1981). *J. Lipid Res.*, **22**, 1235.
21. Holmquist, L., Carlson, K., and Carlson, L. A. (1978). *Anal. Biochem.*, **88**, 457.
22. Swaney, J. B. and Gidez, L. I. (1977). *J. Lipid Res.*, **18**, 69.
23. Cobelli, C., Toffolo, G., Bier, D. M., and Nosadini, R. (1987). *Am. J. Physiol.*, **256**, E551.
24. Cryer, D. R., Matsushima, T., Marsh, J., Yudkoff, M., Coates, P. M., and Cortner, J. A. (1986). *J. Lipid Res.*, **27**, 508.
25. Packard, C. J., Third, J. L. H. C., Shepherd, J., Lorimer, A. R., Morgan, H. G., and Lawrie, T. D. V. (1976). *Metabolism*, **25**, 995.
26. Turner, P. R., Konarska, R., Revill, J., Marara, L., La Ville, A., Jackson, P., Cortese, C., Swan, A. V., and Lewis, B. (1984). *Lancet*, **ii**, 663.
27. Vega, G. L. and Grundy, S. M. (1989). *Atherosclerosis*, **76**, 139.
28. Matthews, C. M. E. (1957). *Phys. Med. Biol.*, **2**, 36.
29. Zilversmit, D. B. (1960). *Am. J. Med.*, **29**, 832.
30. Soutar, A. K., Myant, N. B., and Thompson, G. R. (1982). *Atherosclerosis*, **43**, 217.
31. Sigurdsson, G. (1982). In *Lipoprotein Kinetics and Modelling* (ed. M. Berman, S. M. Grundy, and B. Howard), pp. 113–120. Academic Press Inc., London.
32. Berman, M., Hall, M., Levy, R. I., Eisenberg, S., Bilheimer, D. W., Phair, R. D., and Goebel, R. H. (1978). *J. Lipid. Res.*, **19**, 38.

6

Lipoprotein–receptor interactions

KAY S. ARNOLD, THOMAS L. INNERARITY, ROBERT E. PITAS, and ROBERT W. MAHLEY

1. Introduction

The interaction of lipoproteins with their receptors is the first step of a homeostatic mechanism that controls plasma cholesterol levels, intracellular cholesterol metabolism, and lipoprotein catabolism. The classic studies of M. S. Brown and J. L. Goldstein (Univ. of Texas, Dallas) have demonstrated the significance of low density lipoprotein (LDL) receptors in supplying cells and tissues with cholesterol and in removing plasma lipoproteins via the liver (1). Other less well-defined lipoprotein receptors are the chylomicron remnant receptor (apolipoprotein E receptor), the high density lipoprotein (HDL) receptor, the acetyl LDL receptor (scavenger receptor), and the immunoregulatory receptor. While each of these receptors is unique, many of the procedures developed for the characterization of the LDL receptor can be applied to study other receptors (2–4).

The majority of the methods described in this chapter were developed between 1973 and 1977 in the laboratories of M. S. Brown and J. L. Goldstein (5) and have been used since that time in our laboratories in over 1500 tissue culture experiments (3). Some of these procedures remain essentially unchanged, while others have been modified. It is our hope and purpose that the following experimental procedures are sufficiently detailed so that any investigator can follow the procedures and successfully and proficiently perform the various LDL receptor assays.

Brown and Goldstein first defined the LDL receptor pathway in cultured human fibroblasts, and their studies have served as a model for receptor-mediated endocytosis in a number of different receptor systems. Low density lipoproteins and apolipoprotein (apo) E-containing lipoproteins bind to the amino-terminal ligand-binding region of the transmembrane LDL receptor. The lipoprotein-receptor complexes concentrate in clathrin-coated pits on the plasma membrane. These coated pits then 'pinch off' inside the cell to form coated vesicles that fuse together, lose their clathrin coats, and become endosomes. Hydrogen pumps acidify the endosomes, causing the receptor to release the lipoprotein. The receptor is then recycled to the cell surface, the lipoproteins are delivered to lyosomes where the apoliproteins are degraded to their constituent amino acids, and the cholesteryl esters and triglycerides are hydrolysed. The cholesterol released from the lysosomes is used in the synthesis of cell membranes and, in certain specialized cells, serves as a sterol precursor (1).

An oversupply of lipoprotein-derived cholesterol to cells has three major metabolic effects. First, it reduces the production of cellular cholesterol by turning off the synthesis of the enzyme 3-hydroxy-3-methylglutaryl-CoA reductase (HMG-CoA reductase), the major rate-limiting enzyme in cholesterol biosynthesis. Second, the excess cholesterol activates and serves as a substrate for acyl CoA:cholesterol acyltransferase (ACAT), which re-esterifies the excess cholesterol to cholesteryl esters that are then stored in the cytoplasm as lipid droplets. Third, excess cholesterol down-regulates the biosynthesis and expression of the LDL receptor (1).

Assays for the major steps in the LDL receptor pathway have been developed. Assays for the binding of lipoproteins to the LDL receptor described in this chapter include those for:

(a) LDL receptors on intact cells (Section 3),
(b) LDL receptors on plasma membrane preparations (Section 6),
(c) detergent-solubilized LDL receptors (Section 6), and
(d) LDL receptors from solubilized cellular extracts on SDS–polyacrylamide gel electrophoresis (ligand blots) (Section 7).

2. Preparation of cell cultures

2.1 Cell lines and strains

Cells grown in culture express the LDL receptor to differing degrees and can thus potentially be used in a cell-surface LDL receptor binding assay. Cultured fibroblasts, smooth muscle cells, Chinese hamster ovary (CHO) cells, and liver cell lines have been most commonly used. Cells are subcultured from flasks into individual wells 5 to 7 days prior to performing the receptor binding assay with the goal of obtaining cultures that are near confluence at the time of the experiment. The procedure for subculturing the cells for an experiment is described in *Protocol 1*.

Protocol 1. Procedure for subculturing cells

1. Remove the culture medium from the flask, and wash the cell monolayer once with ~2 ml per T75 flask of trypsin/EDTA (0.05% trypsin, 0.53 mM EDTA, 0.22-μm-filtered and stored in aliquots at 4 °C for ~1 month or −20 °C until use.)

2. Aspirate the trypsin wash from the flask, add another 3–4 ml of trypsin, and lay the flask horizontally.

3. When the cells start to round up (but not detach) in the flask (1 to 3 min), immediately aspirate the trypsin and dislodge the cells by tapping the flask vigorously on its side. (If the cells have started to detach before the trypsin is removed, dislodge the cells by tapping the flask, add medium containing 10% serum, remove all medium (including cells) to a sterile centrifuge tube, centrifuge at 1000 r.p.m. for 10 min, and aspirate the medium from the cells.)

4. Add 10 ml of warm (37 °C) serum-containing culture medium to the cells. (We use Dulbecco's Modified Eagle's Medium (DMEM) containing 10% fetal bovine serum (FBS) for most fibroblast and smooth muscle cell lines.) Suspend the cells thoroughly by repetitive pipetting, and count an aliquot of the cell suspension using a Coulter Counter or haemocytometer.

5. For plating, dilute the cell suspension as described in *Table 1*. A uniform cell suspension is imperative and can be assured by a constant mixing of the contents of the flask during plating. Pipette the suspension quantitatively into each well. Assure an even distribution of cells in the wells by gently shaking the dish back and forth (not swirling) and then in a diagonal direction several times. (Cluster dishes [Falcon Labware], which contain multiple wells molded into one plate, greatly speed subculturing and the receptor binding assays themselves. They are available in 24 × 16 mm, 12 × 22 mm, and 6 × 35 mm sizes.)

Table 1. Representative cell numbers for subculturing[a]

Cell line	Plate size diameter (mm)	Vol. of cell suspension per well (ml)	Cells/well[b]
Human fibroblasts	16	1	9×10^3
	22	1	14×10^3
	35	2	34×10^3
	60	3	60×10^3
	100	5	150×10^3
Dog smooth muscle	22	1	12×10^3
	35	2	30×10^3
	60	3	50×10^3
Mouse peritoneal macrophages	16	1	4×10^5 macrophages/plate
	22	1	1×10^6
	35	2	$2–3 \times 10^6$
	60	3	5×10^6
	100	5	10×10^6

[a] The number of cells needed should be adjusted for differences in cell growth, but this table should serve as an estimate for initial trials.

[b] This table assumes a 7 day growth period. For a shorter growth period, add an additional one-third of these cell numbers for each day less of growth.

6. An alternative method for subculturing cells is based on the split ratio used to maintain stock flasks rather than on a cell count. For example, the

Protocol 1. *Continued*

growth area of one T75 flask is 75 cm^2 and that of a 35 mm dish or well is 9.6 cm^2. The ratio between these is 7.8. Therefore, approximately eight 35 mm dishes are equivalent in area to that of a T75 flask. Based on a split ratio of 1:4 used for human fibroblasts, one confluent T75 flask will yield approximately 32 confluent 35 mm dishes in 1 week. This technique for plating cells is useful when first working out conditions for the growth of a cell line or strain.

7. Grow the cells in a 7% CO_2 incubator for 5 to 7 days, and 24 to 48 h before the experiment, change the medium to DMEM containing 10% lipoprotein-deficient serum (LPDS) to up-regulate the expression of LDL receptors. Wash the cells once with medium containing 5% LPDS before adding medium containing 10% LPDS. Number the wells on the bottom of each dish.

8. On the day of the experiment, the cells should be about 90–95% confluent and evenly distributed over the surface of the dish.

2.2 Mouse peritoneal macrophages

Mouse peritoneal macrophages are obtained by a modification of the procedure of Cohn and Benson (6). We use female Swiss/Webster mice (females can be housed together with less fighting).

Protocol 2. Procedure for obtaining mouse peritoneal macrophages

1. Anesthetize four mice at one time in a closed jar with 2–3 ml of Metofane (methoxyflurane, Pitman-Moore).

2. Sacrifice the anesthetized mice by cervical dislocation, and lay them on their backs. Flood the abdominal area with 95% EtOH, and snip the abdominal skin while tenting it from the peritoneum with forceps. Enlarge the hole by inserting scissors and spreading open the skin.

3. Use a 12 cc syringe with a 20 or 22 gauge needle to inject 8–10 ml of sterile phosphate-buffered saline (PBS) into the abdominal cavity. Agitate the peritoneum by gently shaking the mouse, and withdraw the peritoneal fluid (which contains resident macrophages) with a sterile syringe and needle while holding the mouse on its side. Combine the peritoneal fluid from several mice into 50 ml sterile tubes on ice. Discard any fluid that is bloody or discoloured (discoloured fluid could indicate a nicking of the bowel).

4. Centrifuge at 4 °C, 200g (r_{av}), 15 min, and thoroughly resuspend the cells in 5–10 ml of DMEM. Count the cells on a Coulter Counter, using a Channelyzer (Coulter Electronics) to visualize the cell populations. The peritoneal exudate will contain platelets, red blood cells, larger lympho-

cytes, and a predominant population of macrophages. Make the final dilution based on the number of cells, factoring in the actual percentage of macrophages in the sample. Usually $1-3 \times 10^6$ macrophages are obtained from each mouse.

5. Pipette the cell suspension as described in *Table 1*. Shake the dishes gently back and forth, then diagonally, several times to distribute the cells evenly in the wells. Incubate the dishes for 1 to 1.5 h in a 7% CO_2 incubator at 37 °C to allow the macrophages to attach.

6. Wash off any non-adherent cells with three washes of DMEM, gently shaking the wells during each wash. Add DMEM containing 10% FBS, and incubate overnight. The following day or just before use, shake gently, and wash the cells once with DMEM containing 10% FBS.

2.3 Human monocyte-macrophages

Human monocytes are isolated from blood and allowed to differentiate into macrophages by culturing for 7 days.

Protocol 3. Procedure for obtaining human monocyte-macrophages

1. Collect blood into acid citrate dextrose (ACD): 9 ml of ACD per 50 ml of blood.

2. Transfer to sterile 50 ml conical tubes, and centrifuge at 500g (r_{av}) for 35 min at 4 °C.

3. Aspirate the plasma, and collect the buffy layer of white blood cells that resides between the plasma and the red cells.

4. Aliquot 20 ml of cell suspension per 50 ml tube, and add 15 ml of PBS to each tube. Underlay with 15 ml of Ficoll-Hypaque (Falcon Labware) separating medium (density = $1.077-1.080$ g/ml) as first described by Boyum (7) using a long spinal needle, and centrifuge at room temperature for 45 min at 400g (r_{av}, brake off).

5. Alternatively, dilute whole blood 1:1 with PBS, and layer it directly onto Ficoll-Hypaque (or underlay it with Ficoll-Hypaque). This procedure usually results in better recovery of white cells, but it can also result in a higher percentage of red cell and platelet contamination.

6. Aspirate most of the upper PBS layer. Collect the white blood cell layer at the interface between the Ficoll-Hypaque and PBS layers. Dilute the white blood cells with five volumes of PBS, and pellet the cells by low-speed centrifugation (15 min, 300g (r_{av}), 4 °C).

7. Remove as much of the supernatant as possible, and resuspend the cells in 40 ml of PBS. Pellet the cells by centrifuging 10 min at 4 °C, 200g. Repeat once.

Protocol 3. *Continued*

8. Resuspend the cells in 10 ml of DMEM by repeated pipetting. Take an aliquot for cell counting on the Coulter Counter, using the Channelyzer to determine the percentage of monocytes.

9. Dilute to the desired concentration with DMEM, and plate the cells according to *Table 1*.

10. After 1 to 1.5 h at 37 °C in a 7% CO_2 incubator, wash the cells three times with DMEM, gently shaking to remove non-adherent cells. Culture with DMEM containing 10–20% autologous human serum for approximately 7 days, feeding with fresh medium on day 3.

3. Receptor binding assays, whole cells

3.1 Direct cell surface receptor binding

At 4 °C, lipoproteins bind to LDL receptors, but the lipoprotein-receptor complexes are not internalized; the lipoproteins are in equilibrium with the LDL receptors, and equilibrium binding studies can be performed. The data from these studies can be subjected to Scatchard analysis (8) to determine the affinity of the lipoprotein for the LDL receptor and the number of receptor-bound lipoproteins per cell. The concentrations of the lipoproteins used depend upon the equilibrium dissociation constant (K_d), and roughly equal numbers of concentrations above and below the K_d should be used. The following are commonly used concentration ranges of radiolabelled lipoproteins: human LDL, 0.25–12 µg/ml; human type III β-very low density lipoproteins (βVLDL), 0.2–10 µg/ml; canine apo E HDL_c, 0.01–1.2 µg/ml; canine βVLDL, 0.05–4 µg/ml; and apo E–dimyristoylphosphatidylcholine (DMPC) complexes, 0.005–1.0 µg/ml (all concentrations are based on Lowry *et al.* (9) protein determinations).

Protocol 4. Procedures used to study the equilibrium binding of ^{125}I-labelled lipoproteins to LDL receptors on intact cells

1. Prior to the experiment, pre-cool the culture dishes containing cells for 10–15 min on a metal tray resting on ice in a large dishpan.

2. The lipoprotein solutions, prepared in 15 ml plastic conical tubes and pre-cooled on ice for 15 min, consist of DMEM–Hepes–LPDS (DMEM buffered with 25 mM *N*-2-hydroxyethylpiperazine-*N'*-2-ethanesulphonic acid (Hepes), pH 7.4 (instead of bicarbonate) and containing 10% LPDS) containing ^{125}I-labelled lipoproteins at 6 to 12 different concentrations for duplicate dishes. Solutions for a third dish at each ^{125}I-labelled lipoprotein concentration are used to assess non-specific binding by including a 20- to 100-fold excess of unlabelled ligand.

3. While dishes of cells are still on ice, remove the culture medium from the

cells and replace it with the medium containing the radiolabelled lipoprotein. (We use 0.25 ml for 16-mm, 0.45 ml for 22-mm, 0.95 ml for 35-mm, and 1.9 ml for 60-mm culture dishes.) Take two aliquots (25 μl each) from each tube for the determination of radioactivity. These are used for Scatchard analysis calculations and to verify the concentration of the radioactive lipoprotein. Incubate and gently rock the cells in the cold, on ice, for 3–5 h. (Receptor binding of most lipoproteins reaches equilibrium at 4 °C in 5–6 h, but for most purposes, a 3–4 h incubation approaching equilibrium is sufficient.)

4. At the end of the incubation and while keeping the cells on ice, remove the medium, and wash the cell monolayers three times with cold PBS– bovine serum albumin (PBS–BSA) (PBS containing 2 mg of BSA/ml), followed by two 10 min incubations with the same buffer and two rapid washes using cold PBS. For washing the cells, use a length of tubing attached to a repeating dispenser. The washes are directed down the side of the well to cover the cells. Cells vary greatly in the strength of their attach- ment to the culture plate, so greater care is needed with some cultures.

5. Remove the cells from the ice, and add 0.5 ml of 0.1 M NaOH to each well using a repeating dispenser. Rock the plates gently by hand to ensure coverage of the entire plate with NaOH. Cover, and incubate at room temperature for 10 min to dissolve the cells.

6. Remove the dissolved cells into 12 × 75 mm tubes. Wash each well once with 0.5 ml of NaOH, and add the rinse to the tube.

7. Cork the tubes and count the ^{125}I in a gamma counter. Most counters are now capable of averaging replicates, subtracting background, and calculating the percentage of the 100% control. Find out if your gamma counter is capable of this and how to set it up. It will save you hours of manual calculation.

8. Store the tubes at 4 °C until protein concentrations can be measured. The Technicon Auto Analyzer II apparatus with the Technicon Lowry protocol is used by our laboratories with BSA as a standard. The Auto Analyzer automatically samples directly from the original 12 × 75 mm tube. When using monolayers of cells that are contact inhibited (e.g. fibroblasts) and near confluency, cell proteins are almost identical in every well and thus only representative determinations are needed. However, when using cells that are not contact inhibited (e.g. smooth muscle cells) or cells that do not divide in culture (e.g. macrophages), every tube should be assayed because of the variability among wells.

3.1.1 Scatchard analysis

Data obtained from the equilibrium binding studies can be linearized by Scatchard analysis (8). From this analysis, the affinity constants (K_d) of the

[125]I-labelled ligands for the receptor and the number of receptor-bound lipoproteins at receptor saturation can be determined. By plotting nanograms bound (B) on the *x*-axis and bound/free (B/F) on the *y*-axis, a straight line will be achieved if the ligand and receptors are a homogeneous population at equilibrium and if they interact in a simple bi-molecular reaction (e.g. equilibrium binding of LDL to the LDL receptor). An interactive BASIC program for the IBM PC called 'Scatdata' is available (10) that converts raw c.p.m. data from direct binding experiments into a Scatchard plot, calculating the K_d, B_{max} (maximum bound), and correlation coefficient of the line via linear regression. *Figure 1* shows the results of a typical experiment on a Scatdata graph.

If the plot of B versus B/F is obviously not a straight line, the analysis of the ligand–receptor binding is more complicated. In such cases, the lipoprotein may be heterogeneous or there may be more than one type of receptor (or other binding site) present. Unfortunately, the simple graphic methods such as the Scatchard plot become inappropriate for analysing complex isotherms. To determine the apparent K_d values in these cases, we use a set of computer programs also available for the IBM PC, 'Equilibrium Binding Data Analysis (EBDA)' and 'Ligand' (Biosoft).

3.1.2 Competitive binding assay

In competitive binding experiments, unlabelled lipoproteins compete with ^{125}I-labelled lipoprotein for binding to the cell-surface LDL receptors. This

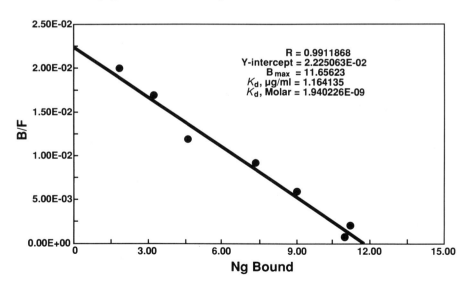

Figure 1. Example of graph produced by 'Scatdata' computer program. This graph shows the results of a typical incubation of human fibroblasts with ^{125}I-labelled human LDL (0.25–25 µg of protein/ml) for 4 h at 4 °C.

procedure is similar to that of the direct binding experiments outlined in *Protocol 4*, with the following differences:

(a) The incubation is on ice for only 2 h.

(b) Use a constant concentration of ^{125}I-labelled lipoprotein (which should approximate its K_d) together with increasing concentrations of unlabelled competing ligand. The unlabelled lipoprotein should be included at concentrations less than and at least 20-fold greater than the K_d of the labelled lipoprotein.

(c) The experiment should include duplicate wells at each concentration of competitor, as well as wells containing the ^{125}I-labelled lipoprotein but not the unlabelled lipoprotein (100% control). Wells that include the ^{125}I-labelled lipoprotein together with a 100-fold excess of unlabelled lipoprotein are used to determine the non-specific binding of the ^{125}I-labelled lipoprotein. Our laboratories routinely use four replicates of the 100% and non-specific controls, two at the beginning and two at the end of the experiment to help correct for any systematic error.

(d) The 50% competition point of a competitor (the lipoprotein concentration required for 50% inhibition of the ^{125}I-labelled lipoprotein binding) can be mathematically determined as follows: convert the competitor concentrations (μg/ml) to a logarithmic value and the percentage bound values to probits. Probit transformation can be performed manually using a published table (11). A few statistical computer graphics programs also do the probit transformation (Plotit, Scientific Programming Enterprises, and Sigma Plot, Jandel Scientific).

With probit-transformed percentage bound data on the *y*-axis and logarithmic concentration on the *x*-axis, a straight line results from which the 50% value can be determined by linear regression. The above-named computer graphics programs have the advantage of performing the transformations, presenting the graph, and determining and printing the 50% values.

3.1.3 Assays that use antibodies in receptor binding studies

Specific polyclonal and monoclonal antibodies are valuable tools for studying lipoprotein–receptor interactions. Specific polyclonal antibodies have been used to demonstrate that an unusual LDL receptor mediates the uptake of βVLDL on mouse peritoneal macrophages (12) and that the LDL receptor-like protein can bind to βVLDL to which exogenous apo E has been added (13). Monoclonal antibodies have been used to map the receptor-binding domain of apo B100 and apo E and to determine which apolipoprotein in a lipoprotein is binding to a receptor. For these types of studies, it is necessary to use high-affinity antibodies that bind to the apolipoprotein or receptor with an affinity similar to or greater than the affinity of the lipoprotein for the receptor.

Protocol 5. Procedure for binding studies using antibodies

1. Purified IgG or Fab fragments should be used for receptor binding assays because endogenous lipoproteins in even small amounts of antiserum or ascites fluid can compete with the ^{125}I-labelled lipoprotein, giving a false-positive result. Because macrophages possess Fc receptors, Fab fragments must be used instead of IgG when studying antibodies on macrophage cultures.

2. If the antibody was made against an apolipoprotein, mix the antibody with the ^{125}I-labelled lipoprotein in DMEM–Hepes–LPDS, and incubate for 1 h at room temperature. Cool the solution, and add it to the chilled cells as described (*Protocol 4*).

3. If the antibody was made against a receptor, dilute the antibody to the desired concentration in DMEM–Hepes–LPDS, and pre-incubate with the cells for 1 h in the cold, on ice. Prepare a 10 × concentrate of the ^{125}I-labelled lipoprotein in DMEM–Hepes–LPDS, add it directly to the plates, and incubate it for an additional 2–3 h on ice. For example, for a final ^{125}I-labelled LDL concentration of 2 μg/ml in 1 ml of medium, add 900 μl of antibody-containing medium for the 1 h pre-incubation followed by 100 μl of a 20 μg/ml solution of ^{125}I-labelled LDL.

3.2 Binding, internalization, and degradation assays

When ^{125}I-labelled lipoproteins are incubated with cultured cells at 37 °C, the lipoproteins are bound by the receptors, internalized, and degraded in the lysosomes. The ^{125}I-labelled apolipoproteins are degraded to small peptides and amino acids, which are secreted into the culture medium. After about 2 h at 37 °C, a steady state is reached, and the amount of ^{125}I-labelled lipoproteins bound to the receptors and contained within the cell is constant. However, degradation products in the medium continue to increase with incubation time. A wide range of incubation times can be used; we routinely use either 5 or 16 h (overnight) incubations.

The 37 °C experiments are similar to the 4 °C experiments (see *Protocol 4* and Section 3.1.2) with the following exceptions:

(a) The apparent binding constants of lipoproteins to the LDL receptor are higher at 37 °C, therefore, higher concentrations of lipoproteins should be used.

(b) The incubation takes place in a 37 °C 7% CO_2 incubator, therefore, Hepes is not included in the medium. Instead, the DMEM is buffered with sodium bicarbonate.

(c) The lipoproteins and medium should be free of bacterial contamination, especially for overnight incubations. Use sterile DMEM–LPDS, and filter the lipoprotein stock solutions through a pre-rinsed syringe filter with cellulose acetate membrane (0.45 μm for LDL and HDL$_c$; 0.8 μm for VLDL and β-VLDL).

(d) The medium containing the lipoproteins should be at or near 37 °C before adding it to the cells.

(e) After incubation, the cells are chilled on a metal tray in a pan of ice. The incubation medium is removed and saved for the degradation assay (*Protocol 7*), and the cells are washed (*Protocol 4*).

3.2.1 Release of bound lipoprotein

Cells incubated with ^{125}I-labelled lipoproteins at 37 °C contain both surface-bound and internalized lipoproteins. To differentiate them, the surface-bound lipoproteins are released from the cells before the cells are solubilized with 0.1 M NaOH.

Protocol 6. Measurement of surface-bound and internalized lipoproteins

1. Following incubation, chill the cells (15 min on ice), and remove the incubation medium for the degradation assay.

2. Rapidly wash the cells three times with cold PBS–BSA followed by two 10 min PBS–BSA incubations on ice.

3. Add a lipoprotein dissociation reagent, and incubate and gently rock the cells on ice in the cold for 1 h.

   • LDL can be released from cell-surface receptors (4) using 4 mg/ml dextran sulphate or 10 mg/ml heparin in PBS.

   • Lipoproteins that bind with a higher affinity than LDL, such as apo E-containing lipoproteins, can be released using 30 mg/ml of sodium phosphate glass in PBS (sodium hexametaphosphate, 9-V030, J. T. Baker or Phosphate Glass, P-8510, Sigma).

4. Remove a known aliquot into a 12 × 75 mm tube for gamma counting.

5. Aspirate the remaining solution, and wash the monolayer twice with PBS. Add 0.1 M NaOH and continue as in *Protocol 4*.

3.2.2 Degradation assay

After binding to the LDL receptors, the ^{125}I-labelled lipoproteins are internalized and degraded to amino acids in the lysosomes. The degradation assay measures the ^{125}I-labelled tyrosine released into the tissue culture medium. The principle of the assay is to remove the remaining intact [125]I-

155

labelled lipoproteins from the medium by trichloroacetic acid (TCA) precipitation, to remove any free ^{125}I, and to measure the radioactivity of the ^{125}I-labelled tyrosine and small peptides. This method is a modification of the procedure described by Goldstein *et al.* (4).

Protocol 7. Measurement of lipoprotein degradation

1. Pre-number two sets of 12 × 75 mm and one set of 13 × 100 mm glass tubes.

2. To one set of 12 × 75 mm tubes, add PBS–BSA so that the final volume, including the incubation medium from the cells, will be 1 ml. After the cells have been placed on ice, remove the medium to the appropriately numbered 12 × 75 mm tube containing the PBS–BSA. Wash the cell monolayers, and count as described in *Protocol 4.*

3. Use a repeating dispenser to add 200 μl of 50% TCA into each tube containing 1 ml of medium–PBS–BSA. Vortex, and incubate at 4 °C for 30 min.

4. Centrifuge at 2000g (r_{av}), 4 °C, 15 min to pellet the TCA precipitate. Remove 1.0 ml of the supernatant into the appropriately numbered 13 × 100 mm tube. Use a repeating pipette to add 10 μl of 40% KI and 40 μl of 30% H_2O_2 to each tube. Vortex, and incubate at room temperature for 5–10 min.

5. Add 2 ml of $CHCl_3$ to each tube and vortex thoroughly (upper aqueous phase should be light pink to light yellow, not dark brown; if dark brown, vortex again). The H_2O_2 converts ^{125}I]iodide to ^{125}I]iodine, which is extracted by the choloroform. Centrifuge for 5 min at 100–200g to separate the aqueous phase from the choloroform phase. Pipette 625 μl of the upper aqueous phase into the second set of 12 × 75 mm tubes, cork, and count radioactivity. The total amount degraded = c.p.m. × 2.

---

4. Assay of cholesteryl ester formation in cultured cells

Cells store excess cholesterol as cholesteryl ester. If [14C]oleate is added to the culture medium, the determination of cholesteryl [14C]oleate is an accurate indication of cholesteryl ester formation. Our procedure is a modification of one from Goldstein *et al.* (4).

Protocol 8. Measurement of cholesterol esterification

1. Incubate the cells at 37 °C for 6–16 h with DMEM containing the lipoprotein(s) of interest and sodium [14C]oleate–albumin complex (14)

(1.5–2.5 mg of sodium oleate/ml and 0.1 mg of fatty-acid-free albumin/ml, 20 000–35 000 d.p.m. of ^{14}C/nmol of oleic acid). Because serum and LPDS promote cholesterol efflux from cells, more cholesteryl ester formation can be measured when cells are incubated in DMEM without LPDS or serum.

(a) For fibroblasts and smooth muscle cells, a final concentration of 0.1 mM sodium [^{14}C]oleate–albumin complex is used, and for macrophages, 0.2 mM.

(b) The lipoprotein concentrations employed for degradation experiments (see Section 3.2) are also effective for cholesteryl ester synthesis experiments. If ^{125}I-labelled lipoproteins are used, degradation can be assayed on the identical cells used for cholesteryl ester synthesis. Any contamination of the lipids by the ^{125}I-labelled proteins is resolved during TLC; the proteins remain at the origin.

(c) Also included are two to four wells for the 'no lipoprotein' background controls, which will receive medium and sodium [^{14}C]oleate–albumin but no lipoproteins.

2. At the end of the incubation, place the cells on metal trays in pans of ice, aspirate the medium or save it for the degradation assay, and wash the monolayers with PBS–BSA and PBS as described in *Protocol 4*.

3. Extract each well with 3:2 hexane:isopropanol containing a [^{3}H]cholesteryl oleate internal standard. (The [^{3}H]cholesteryl oleate internal standard should contain approximately 100 000 d.p.m. of ^{3}H and approximately 20 μg of unlabelled cholesteryl oleate per ml in 3:2 hexane:isopropanol.) The internal standard d.p.m. should slightly exceed the maximum ^{14}C d.p.m. expected at the highest lipoprotein concentration (50 000–75 000 d.p.m. per well as an initial approximation). If more volume is needed, add 3:2 hexane:isopropanol. For the 100% values, dispense the same amount of [^{3}H]cholesteryl oleate internal standard directly into three scintillation vials, allowing the solvent to evaporate in a hood. At room temperature, perform a 30 min extraction of the cell mono-layers while the dishes are on a rotating platform or rocker; put lids on the dishes to reduce evaporation. After the extraction, remove the solvent from each well into 12 × 75 mm tubes, and rinse the wells once with an equal amount of 3:2 hexane:isopropanol. (The tubes can be covered with foil and stored at −20 °C until needed: Step 5 below.)

4. After the monolayers have dried for a few minutes, dissolve them in 0.1 N NaOH as described in *Protocol 4*.

5. Evaporate the solvent extracts containing cellular cholesteryl [^{14}C]oleate and [^{3}H]cholesteryl oleate internal standard to dryness at 37 °C under a stream of nitrogen.

Protocol 8. *Continued*

6. Dissolve the lipids in 50 µl of 2:1 chloroform:methanol and spot onto channeled, pre-absorbent, silica gel TLC plates (Whatman LK5D, Whatman Inc.). A lane at one end of the plate can be used to spot a neutral lipid mix or cholesteryl oleate solution as standards. However, the mass of unlabelled cholesteryl oleate added to the internal standard should be easily visualized.

7. Develop plates in a pre-equilibrated, filter paper-lined chromatography tank containing 100 ml of 90:10:1 hexane:diethyl ether:ammonia. (Racks are available that allow development of six plates at a time.) Develop for 25–30 min at room temperature or until the solvent front is 2–4 cm from the top of the plate. Air dry under a fume hood.

8. Visualize the lipid bands with idione vapour in a chromatography tank containing iodine crystals by placing the bottom of the plates into a trough so that they are not in direct contact with the crystals. Visualization takes 5–10 min.

9. In this solvent system, the cholesteryl ester band is the uppermost band, running just below the solvent front. Mark the band by scoring the silica gel with a razor blade, and allow the iodine to volatilize in a fume hood. Scrape off the silica gel above the cholesteryl ester bands with a razor blade. Then scrape each cholesteryl ester band onto weighing paper, and pour it into a numbered scintillation vial.

10. Add non-aqueous scintillation cocktail and count radioactivity using a program for dual-labelled samples. Also count the vials that contain only the ^{3}H-labelled internal standard.

11. The amount of cholesteryl ester formed is calculated as follows: [a]

$$\text{pmol of cholesteryl [}^{14}\text{C]oleate} = \frac{\text{d.p.m. of }^{14}\text{C in sample} \times \left[\dfrac{\text{d.p.m. of 100\% [}^3\text{H]cholesteryl oleate}}{\text{d.p.m. of }^3\text{H in sample}}\right]}{\text{specific activity of sodium [}^{14}\text{C]oleate in d.p.m./pmol}}$$

$$\frac{\text{pmol of cholesteryl ester}}{\text{µg of cell protein}} = \frac{\text{pmol of cholesteryl [}^{14}\text{C]oleate}}{\text{µg of cellular protein per dish}}$$

[a] Data handling is greatly expedited if the formula(s) can be stored in a calculator or computer program such that only each sample's ^{3}H and ^{14}C d.p.m. value need be entered, with the µg of cell protein value if available.

5. Cholesterol, cholesteryl ester, and triacylglyceride mass determination

We use a single procedure to measure the mass of cholesterol, cholesteryl ester, and triacylglycerol in a 35 mm well of cells. In our procedure, internal lipid standards are added directly to the organic solvents used to extract the lipids from the cells. Thus, the losses from all extraction and analysis steps can be calculated directly. These neutral lipids are separated by TLC and quantitated by GLC.

Protocol 9. Determination of cholesterol, cholesteryl ester, and triacylglyceride mass

1. The internal standards of stigmasterol, stigmasteryl oleate, and tri-heptadecanoin are used for cholesterol, cholesteryl oleate, and tri-glyceride analysis, respectively. Stigmasterol and triheptadecanoin are available from Supelco and NuChek Prep, Inc., respectively. Stig-masteryl oleate is synthesized by the procedure of Patel *et al.* (15).

2. Incubate the cells with lipoproteins at 37 °C for 24 h (see Section 3.2), and wash as described in *Protocol 4*.

3. For 35 mm wells, add 1 ml of 3:2 hexane:isopropanol containing 5 μg of stigmasterol, 7.5 μg of stigmasteryl oleate, and 5 μg of triheptadecanoin to the cells, and incubate at room temperature for 30 min with gentle mixing on a rocking or rotating platform.

4. Remove the 3:2 hexane:isopropanol from the cells into tubes pre-rinsed with 3:2 hexane:isopropanol. Wash the cells twice with hexane:isopro-panol (no added internal standards), and add the washes to the tubes.

5. Evaporate the samples to dryness under a stream of dry nitrogen, redissolve in 25–50 μl of 2:1 chloroform:methanol, and spot onto channeled TLC plates (see *Protocol 8*, step 6). A neutral lipid mixture is spotted in one lane as a reference.

6. Develop the plate in a chromatography tank pre-equilibrated with 80:20:1 hexane:diethyl ether:concentrated NH_4OH, and remove it when the solvent front is 1–2 cm from the top of the plate.

7. Dry the TLC plate under a hood for several minutes, and then expose it to iodine vapours for visualization of the lipids. Mark the areas corresponding to cholesterol, cholesteryl oleate, and triacylglycerol by scratching with a razor blade or needle. (Allow the iodine to sublime in a fume hood before continuing.)

8. Scrape the cholesterol band, which also contains the stigmasterol internal standard, into pre-rinsed screw-capped (Teflon-lined caps)

Protocol 9. *Continued*

    tubes, and add 1 ml of 2:1 chloroform:methanol. Vortex the mixture, centrifuge it at low speed, and store the supernatant in pre-rinsed tubes.

9. Scrape the triacylglycerol band, which also contains the triheptadecanoin internal standard, into pre-rinsed screw-capped tubes, and transesterify by heating at 70 °C for 2 h in 0.5 ml methanol containing 7% anhydrous HCl. Cool the extract, and add 0.5 ml H_2O and 1 ml hexane. Mix the solvents, and centrifuge the solution at low speed to separate the phases. Remove the upper hexane layer to a pre-rinsed screw-capped tube.

10. Scrape the cholesteryl ester band, containing the stigmasteryl oleate internal standard, into pre-rinsed screw-capped tubes. Hydrolyse the cholesteryl ester in 0.5 ml of methanolic KOH (1 M KOH in methanol) by heating it at 80 °C for 30 min. Extract as described in *Protocol 8*, step 9.

11. These isolated lipids can now be analysed by GLC. Evaporate the extracts to dryness under nitrogen. Add a few drops of iso-octane or chloroform to the tubes to dissolve the lipids.

12. GLC conditions: Analyse free cholesterol and cholesterol resulting from the hydrolysis of cholesteryl esters on a 6 ft column packed with 3% OV-17 on 100/120 mesh Gas Chrom Q (Supelco) at 275 °C. The flow rate for the nitrogen carrier is 33 ml/min. Quantitation is by reference to the stigmasterol internal standard. Analyse the fatty acid methyl esters from the triacylglycerols on a 6 ft column packed with 10% Sp-2330 on 100/120 mesh Chromasorb WAW (Supelco) at 165 °C. The nitrogen gas carrier flow rate is 30 ml/min. Quantitation is by reference to the methyl heptadecanoate (esterified from the triheptadecanoin) internal standard.

6. Membrane binding assay

Receptors can be assayed on crude membranes or on detergent-solubilized crude membranes. We use a modification of the procedures of Basu *et al.* (16) and Schneider *et al.* (17, 18).

Protocol 10. Preparation of membranes from cultured cells

1. Cells in large (100-mm) dishes, subcultured as described in *Protocol 1*, are chilled on a metal tray in a tub of ice. Wash twice with 4–5 ml of cold PBS containing 0.2% BSA, then once with 5 ml of cold PBS.

2. Scrape the cells from the dish using a rubber scraper or plastic cell scraper, tilt the plate, and collect the cells. Rinse the dishes with 1–2 ml of cold PBS, and collect the scraped cells in 50 ml plastic centrifuge tubes.

3. Centrifuge at 4 °C for 5 min at 800g. Aspirate off the s
   the cell pellet, which may be frozen at −76 °C for late
4. Add to each cell pellet approximately 1 ml of cold 10 m
   mM NaCl, 1 mM CaCl₂, pH 7.4. Resuspend each pel
   combine the contents of the tubes.
5. Disrupt the cells with a Polytron homogenizer (Brinkmar
   using a small probe at setting #8; homogenize for 10 sec, ch
   min, and homogenize again for 10 sec.
6. Centrifuge the homogenate at 4 °C for 5 min at 800g.
   supernatant to 38.5 ml polyallomer tubes, and centrifuge in
   L8-70 using a 60 Ti rotor for 60 min at 31 000 r.p.m. (100 000
7. After centrifugation, cut the tube with a tube slicer approximat
   from the bottom. Decant, and discard the supernatant. Layer t
   tubes containing the pellets into 50-ml plastic centrifuge tubes
   three pellets can be stacked into one centrifuge tube), and s
   −76 °C.

Protocol 11. Preparation of membranes from liver

1. Remove the liver, and place it into ice-cold 0.15 M NaCl (saline).
   subsequent steps are also carried out at 4 °C.
2. Homogenize 1 g of liver with two 10 sec pulses in a Polytron homogenizer
   (setting #10) in 10 ml of buffer containing 10 mM Tris−HCl, 150 mM
   NaCl, 1 mM CaCl₂, and 0.5 mM phenylmethylsulphonyl fluoride
   (PMSF), pH 7.5 (Buffer A).
3. Centrifuge the homogenized liver at 500g (r_{av}) for 5 min, save the
   supernatant, and recentrifuge it at 8000g (r_{av}) for 15 min. Discard the
   pellet.
4. Centrifuge the 8000g supernatant at 100 000g (r_{av}) for 60 min at 4 °C, and
   resuspend each 100 000g pellet in 6 ml of Buffer A by passing through a
   22 gauge needle 10 times.
5. Re-sediment this suspension for 60 min at 100 000g.
6. Assay the final pellets immediately or freeze them in liquid nitrogen and
   store them at −80 °C for up to 2 weeks.

6.1 Airfuge membrane assay

Airfuge membrane assay is carried out as described in *Protocol 12.*

Protocol 12. Pr␣␣␣dure for airfuge membrane assay

1. Suspend m␣␣␣nes in ice-cold Buffer B (20 mM Tris–HCl, pH 7.5,
   1 mM Ca␣in concentration by the Lowry method (9).
   Determir␣qual volume of modified Buffer (80 mM Tris–HCl, pH

2. Dilute ␣l₂, 40 mg of BSA/ml) and then to a range of 2–6 mg of
   7.5, 1n Buffer C (equal volumes of Buffer B and modified
   prote

3. Bu␣poproteins in Buffer C should be at a concentration five
   12␣sired final incubation concentration. Final lipoprotein
   ␣s should be the same as those for 4 °C receptor binding
   DL = 0.2–25 µg/ml and apo E HDL$_c$ = 0.01–1.8 µg/ml; see

   ml microcentrifuge tubes:

   Buffer C

   ^{125}I-labelled lipoprotein solution

   membranes

   ipoprotein binding to the LDL receptor requires calcium, excess
   ␣bolishes the lipoprotein–receptor interaction. Therefore, the
   of 20 µl of buffer containing 100 mM EDTA to the incubation
   place of normal Buffer C serves as a control for non-specific

   the mixture, and incubate it on ice for 60 min. During the
   ion, place 100 µl of FBS into numbered airfuge tubes (Beckman
   ␣, Beckman Instruments).

   ␣ the incubation period, vortex the tubes, and layer 75 µl of
   ␣ubation mixture onto the FBS in the airfuge tubes. Centrifuge 30 min
   ␣ 30 p.s.i. (135 000g, r_{av}) with the airfuge at 4 °C.

   Remove the supernatant with a syringe or microlitre pipette with narrow
   tip, and add 175 µl of FBS to the tube without disturbing the pellet.
   Recentrifuge for 10 min at 30 p.s.i.

8. Remove the supernatant, and cut the airfuge tube with a razor blade just
   above the pellet. Place the bottom of the tube with the pellet into a 12 ×
   75 mm tube, cap, and count on a gamma counter. If an airfuge is not
   available, a Beckman TL100 tabletope ultracentrifuge with a TL100.2
   rotor can also be used.

6.2 Soluble receptor assay

The soluble receptor assay is used for receptors solubilized in detergent and can therefore be used as an assay for receptor purification. The procedure involves precipitating the receptors from the detergent-containing solution using acetone/phosphatidylcholine (PC), incubating the receptors with ^{125}I-labelled lipoproteins, and trapping the lipoprotein–receptor complex on cellulose acetate membranes.

Protocol 13. Procedure for soluble receptor assay

1. In 17×100 mm polypropylene tubes with caps, mix:
   - 2.7 ml of PC diluting buffer (50 mM Tris maleate, pH 7.0, 2 mM $CaCl_2$, and 50 mM PMSF added immediately before use).
   - 1.5 ml of 2 mg/ml PC solution (mix well during additional pipetting). To prepare the 2 mg/ml PC solution, add 200 mg (10 ml) of egg PC (Egg Lecithin, Avanti Polar Lipids) to a 100 ml round-bottom flask, and evaporate to dryness under nitrogen. Dissolve in 3–5 ml of diethyl ether, and evaporate to dryness under nitrogen. Add 100 ml of PC diluting buffer, and shake vigorously for 5 min. The PC solution is stored at 4 °C and should be used within 2–3 weeks.
   - 0.5 ml of receptor sample (e.g., column fraction aliquots).

2. While vortexing, add 3.3 ml of −20 °C acetone (Ultrex grade). Centrifuge at 20 000*g* for 20 min at 0 °C (12 000 r.p.m. in JA14 rotor, Beckman J2-21 centrifuge).

3. Aspirate the supernatant, and resuspend the pellet in 240 μl of Buffer C-BSA (50 mM Tris, pH 7.5, 25 mM NaCl, 1 mM $CaCl_2$, 20 mg of BSA/ml).

4. Mix together in 0.5 or 1.5 ml microcentrifuge tubes:
   - 25 μl of Buffer C-BSA or Buffer C-BSA containing 100 mM EDTA
   - 25 μl of ^{125}I-labelled ligand solution
   - 75 μl of sample (mixed well)

   Filter the ^{125}I-labelled lipoprotein at five-times final incubation concentration in Buffer C–BSA through 0.45 μm cellulose acetate membrane syringe filters. Use final lipoprotein concentrations as given in *Protocol 12*. Incubate the mixture on ice for at least 1 h.

5. During this incubation, prepare cellulose acetate membranes to fit your vacuum manifold (0.45 μm pore diameter, Oxoid Ltd). These are single-sided membranes, so be sure to keep the correct side up throughout the procedure. Pre-soak the membranes individually in 3 ml of wash buffer (20 mM Tris, pH 7.5, 1 mM $CaCl_2$, 50 mM NaCl, 1 mg of BSA/ml) using

Protocol 13. *Continued*

35 mm tissue culture cluster dishes. Using forceps, place the filter right side up on the vacuum manifold, and wash with 3 ml of cold wash buffer. Turn off the vacuum, and add 3 ml of wash buffer to each membrane.

6. Mix incubation tubes, and aliquot 90 μl into the buffer on the filter. Pull the buffer through the filter with the vacuum, and wash each filter four times with 3 ml of wash buffer.

7. Remove the membrane using fine forceps, and place it into a 12 × 75 mm tube for gamma counting.

---

7. Ligand blotting

In 1983, Daniel *et al.* (19) demonstrated that after SDS polyacrylamide gel electrophoresis, the LDL receptor could be transferred to nitrocellulose paper and detected by ligand blotting. Since then, this technique has been widely used (and abused) to identify receptors in crude membrane extracts. In ligand blotting, the isolated receptor is electrophoresed on SDS polyacrylamide gels and then electrophoretically transferred to nitrocellulose paper. Incubation of the nitrocellulose blot in a solution containing a ligand of the receptor results in the ligand binding to the receptor on the blot. If the ligand is isotopically labelled, the receptor can be localized by autoradiography.

Ligand blotting offers some information not available from receptor binding assays:

(a) determination of the receptor's molecular weight,

(b) identification of receptors for which no anti-receptor antibody is available, and

(c) confirmation that the ligand is interacting with a specific cell-surface molecule.

The interpretation of ligand blots requires great care because artifactual conclusions can easily be drawn. For example, the technique is only valid if the receptor retains its ligand-binding capacity after electrophoresis in SDS. The second major problem is that the ligand may bind to a protein that is not the physiological receptor but nevertheless interacts with the ligand on ligand blots. Finally, if the ligand is 'sticky', it may bind strongly to the nitrocellulose, causing a high background and obliterating any specific binding. The procedure we follow is a modification of that of Daniel *et al.* (19).

Protocol 14. Procedure for ligand blotting

1. The gel electrophoresis sample application buffer should contain no reducing agent, and the samples are not heated.

2. After electrophoretic transfer (20), stain the nitrocellulose blot for a few minutes in Ponceau S, a water-soluble dye that stains the proteins red so that they can be temporarily visualized (and even photocopied), but washes off during subsequent steps.

3. To minimize non-specific protein binding to the nitrocellulose sheet, incubate in 5% (w/v) blotto buffer (Carnation nonfat dry milk in 50 mM Tris, 2 mM $CaCl_2$, 80 mM NaCl, pH 8.0) for 60 min at room temperature or overnight at 4 °C.

4. Remove the blotto buffer, and replace with fresh buffer containing the ^{125}I-labelled ligand (5–10 μg/ml). (Heat-sealable plastic bags make excellent, easily disposable containers for the blot and solutions.)

5. Incubate the blot for 2–4 h at room temperature or overnight at 4 °C on a rocking or rotating platform.

6. After the incubation, rinse the blot three times followed by two 10-min washes with blotto buffer. Rinse it with distilled water, air dry, and autoradiograph with an intensifying screen at −70 °C.

8. Fluorescent probes

Recent developments for studying receptor–ligand interactions make use of fluorescently labelled ligands, the Fluorescence-Activated Cell Sorter (FACS), and the confocal microscope with video imaging analysis.

8.1 DiI

DiI (1,1'-dioctadecyl-3,3,3',3'-tetramethylindocarbocyanine) is an attractive probe because it is highly fluorescent, commercially available, inexpensive, and easily incorporated into lipoproteins. With an excitation wavelength of 520 nm and emission wavelength of 570 nm, DiI can be visualized using rhodamine settings or filters. Once a DiI-labelled lipoprotein is internalized, the DiI is retained in the cell for several days, providing a stable marker of lipoprotein uptake.

Protocol 15. Procedure for the incorporation of DiI into lipoproteins

1. Dissolve DiI (Molecular Probes, Inc., catalogue #D282) in dimethyl sulphoxide (DMSO) at a concentration of 3 mg/ml. Filter 2 ml of LPDS/ mg of lipoprotein together with the lipoprotein into a sterile tube

Protocol 15. *Continued*

containing 50 μl of DiI/mg of lipoprotein. (Use a pre-rinsed 0.45 μm syringe filter for LDL or HDL$_c$ and a 0.8-μm filter for the larger β-VLDL or VLDL.) Wrap foil around the tube to shield it from light, and incubate the solution at 37 °C for 8 h to overnight.

2. Re-isolate the lipoprotein by density gradient ultracentrifugation:

    (a) Raise LDL or HDL$_c$ to $d = 1.063$ g/ml and VLDL or βVLDL to $d = 1.02$ g/ml with solid KBr.

    (b) Centrifuge at 165 500g (r_{av}), 4 °C, 16–18 h (e.g. 50 000 r.p.m., 50 Ti rotor, Beckman L8-70 centrifuge).

    (c) Isolate the DiI-labelled lipoproteins from the top of tube (you can use tube slicing).

3. Dialyse at 4 °C against 0.15 M NaCl, 0.01% EDTA in a foil-covered container so as to prevent bleaching of the DiI. (It is important to note that ^{125}I-labelling and reductive methylation should be performed on the lipoprotein before DiI labelling, whereas acetoacetylation of the lipoprotein should be done after DiI incorporation.)

4. Incubate DiI-labelled lipoproteins with cells at either 4 °C or 37 °C as described in *Protocol 4* and Section 3.2.

The following are some uses for DiI-labelled cells:

(a) Cells can be viewed on an inverted microscope and then replaced in culture (e.g. after isolating individual cells or clones that have taken up the DiI-labelled lipoproteins). Wash the cells three times with culture medium under sterile conditions, and add fresh culture medium before viewing.

(b) Permanent slides of DiI-labelled cells can be prepared. Use tissue culture chamber slides (LabTek Products) to plate cells that will be fixed; the chamber that holds the culture medium can be removed after fixation to yield a flat side. Wash the cells as in *Protocol 4*, and fix them with 4% Formalin in PBS. Incubate with fixative for 30 min, rinse with PBS, and mount using aqueous mounting medium (Gelf Mount, Biomeda Corp, catalogue #M01). Cover with a coverslip, and store away from direct light. The coverslip can be sealed with clear nail polish. For black-and-white photography of DiI-labelled cells, use Ilford HP05 film (Ilford Photo Corp).

(c) The amount of lipoprotein taken up and degraded by cells can be quantitated using DiI-labelled lipoproteins. First, the amount of DiI in the lipoprotein is determined by comparing the fluorescence of the DiI-labelled lipoprotein dissolved in methanol with a standard curve of DiI powder dissolved in methanol. The DiI fluorescence of cells extracted with methanol compared to the DiI fluorescence of the lipoproteins provides an accurate measure of the cellular uptake of the lipoproteins. The excitation and emission wavelengths are 520 and 570 nm, respectively.

(e) FACS: To prepare cells for fluorescence-activated cell sorting using DiI-labelled lipoproteins, culture the cells as in *Protocol 1*, pre-treat them if desired, and incubate them with or without the fluorescently labelled (DiI) lipoprotein for 5–16 h at 37 °C. Determine the background in cells incubated with non-DiI-labelled lipoproteins. Wash the cells with culture medium containing 10% serum or LPDS, and remove them from the plate by trypsinizing or scraping. Pellet the cells and resuspend them in PBS by repeatedly pipetting them and passing them through a 10-μm nylon mesh syringe filter to obtain a single-cell suspension. Collecting and filtering should be performed immediately before FACS sorting. *Figure 2* shows the results of a FACS analysis of human fibroblasts incubated with DiI–LDL.

8.2 DiO

DiO (3,3′-dioctadecyloxacarbocyanine perchlorate) is another fluorescent probe (Molecular Probes, catalogue #D275) that is easily incorporated into lipoproteins. Its excitation and emission wavelengths (484 nm and 507 nm, respectively) are similar to those of fluorescein, thus providing the opportunity for a dual-label study when used with DiI. It is incorporated into the lipoprotein using the same procedure as outlined for DiI (*Protocol 15*). However, since there is more background fluorescence with DiO than with

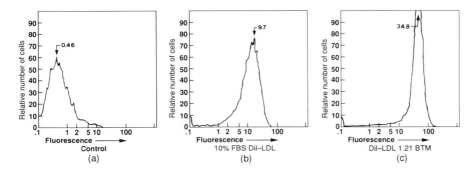

Figure 2. Fluorescence-activated cell sorting of human fibroblasts labelled with DiI–LDL. (a) Control fibroblasts have LDL (no DiI) added and serve as a control for cellular and lipoprotein autofluorescence. Peak fluorescence intensity = 0.46. (b) Fibroblasts grown in the presence of 10% fetal bovine serum (FBS). Because FBS provides most of the cholesterol that the fibroblasts need for growth, the LDL receptors are down-regulated under these conditions. Peak fluorescence intensity = 9.7. (c) Fibroblasts grown in the presence of 10% LPDS (d = 1.21 g/ml bottom fraction [1.21 BTM]) for 48 h show greatly up-regulated LDL receptors. Peak fluorescence intensity = 34.8.

DiI, DiI is the probe of choice for single-label experiments. If dual-label is desired, DiO is an available second probe.

References

1. Brown, M. S. and Goldstein, J. L. (1986). *Science*, **232**, 34.
2. Mahley, R. W. and Innerarity, T. L. (1983). *Biochim. Biophys. Acta*, **737**, 197.
3. Innerarity, T. L., Pitas, R. E., and Mahley, R. W. (1986). *Methods Enzymol.*, **129**, 542.
4. Goldstein, J. L., Basu, S. K., and Brown, M. S. (1983). *Methods Enzymol.*, **98**, 241.
5. Goldstein, J. L. and Brown, M. S. (1977). *Ann. Rev. Biochem.*, **46**, 897.
6. Cohn, Z. A. and Benson, B. (1965). *J. Exp. Med.*, **121**, 153.
7. Boyum, A. (1976). *Scand. J. Immunol.*, **5**, 9.
8. Scatchard, G. (1949). *Ann. N.Y. Acad. Sci.*, **51**, 660.
9. Lowry, O. H., Rosebrough, N. J., Farr, A. L., and Randall, R. J. (1951). *J. Biol. Chem.*, **193**, 265.
10. Send a formatted diskette, 5¼″ or 3½″, to Kay Arnold, Gladstone Foundation Laboratories, P.O. Box 40608, San Francisco, CA 94140.
11. Fisher, R. A. and Yates, F. (1953). In *Statistical Tables for Biological, Agricultural and Medical Research*, 4th Edn, p. 60. Oliver and Boyd, Edinburgh.
12. Koo, C., Wernette-Hammond, M. E., and Innerarity, T. L. (1986). *J. Biol. Chem.*, **261**, 11194.
13. Kowal, R. C., Herz, J., Goldstein, J. L., Esser, V. and Brown, M. S. (1989). *Proc Natl. Acad. Sci. USA*, **86**, 5810.
14. Van Harken, D. R., Dixon, C. W. and Heimberg, M. (1969). *J. Biol. Chem.*, **244**, 2278.
15. Patel, K. M., Sklar, L. A., Currie, R., Pownall, H. J., Morrisett, J. D., and Sparrow, J. T. (1979). *Lipids*, **14**, 816.
16. Basu, S. K., Goldstein, J. L., and Brown, M. S. (1978). *J. Biol. Chem.*, **253**, 3852.
17. Schneider, W. J., Basu, S. K., McPhaul, M. J., Goldstein, J. L., and Brown, M. S. (1979). *Proc. Natl. Acad. Sci. USA*, **76**, 5577.
18. Schneider, W. J., Goldstein, J. L., and Brown, M. S. (1980). *J. Biol. Chem.*, **255**, 11442.
19. Daniel, T. O., Schneider, W. J., Goldstein, J. L., and Brown, M. S. (1983). *J. Biol Chem.*, **258**, 4606.
20. Burnette, W.N. (1981). *Anal. Biochem.*, **112**, 195.

Assay of lipoprotein lipase and hepatic lipase

GUNILLA BENGTSSON-OLIVECRONA and THOMAS
OLIVECRONA

1. Introduction

Lipoprotein lipase (LPL) and hepatic lipase (HL) are two key enzymes in the metabolism and interconversion of lipoproteins (1–3; Table 1). The reactions occur at endothelial surfaces of blood vessels, where the enzymes are attached by interaction with polyanions, probably heparan sulphate proteoglycans anchored in the endothelial cell membrane (4). LPL appears all over the vascular bed of most extrahepatic tissues, though in varying amounts, while HL is localized at the liver and some steroid hormone producing glands. LPL acts mainly on the triglycerides of chylomicra and VLDL. The released fatty acids are taken up by nearby cells or are circulated in blood as albumin-bound free fatty acids. LPL is under metabolic and hormonal control. HL acts mainly on smaller lipoproteins (HDL, IDL, chylomicron remnants). Its function is not yet fully understood. Some animal species, e.g. guinea pigs and cows, have little or no HL. HL is regulated by steroid hormones, but does not seem to be metabolically regulated.

The primary structures of LPL and HL from several species are known

Table 1. Properties of lipoprotein lipase (LPL) and hepatic lipase (HL)

	LPL	HL
Substrate specificity	Same lipids, same positional specificity	
Preferred substrate	Chylomicra	HDL
	VLDL	Remnants
Tissue localization	Adipose tissue	Liver
	Heart, muscles etc.	Steroid-hormone producing glands
Regulation	Rapid, feeding–fasting, exercise	Slower, endocrine
Half-life *in vivo* (rats)	< 1 h	> 4 h
Species distribution	Activity always high	Variable

from cloning of their cDNAs. The enzymes show strong homologies to each other and also to pancreatic lipase, indicating that they have evolved from a common ancestral gene (5). A corollary is that the enzymes probably have similar active sites. The catalytic properties of LPL and HL are in fact similar in most respects. They hydrolyse tri-, di-, and mono-glycerides, phospholipids, and a variety of model substrates, e.g. paranitrophenyl esters. In their action on glycerides LPL and HL show specificity for the sn-1,3 ester bonds, with a preference for the sn-1 ester bond. They do not show any marked preference for specific fatty acids, except that LPL, but not HL, hydrolyses ester bonds involving long polyunsaturated fatty acids more slowly than ester bonds involving other fatty acids (6).

The importance of the lipases for lipid transport and lipoprotein inter-conversions is demonstrated by the major derangements of lipoprotein metabolism which occur in individuals with genetic defects (1). There are many reports on correlations between lipase activities and lipoprotein or disease parameters in clinical and population studies (1, 7, 8). Therefore, measurement of lipase activities should be considered in studies on the effects of diet, drugs or other parameters on lipoprotein metabolism.

Cloned cDNAs for both lipases from several species are available in some laboratories, and can be used to estimate mRNA levels by standard techniques. A general finding is that the metabolic regulation of LPL in adipose tissue and muscles is not fully explained by changes in its mRNA, but other control mechanisms must also operate.

Antisera to LPL and HL are available in some laboratories, and can be used for immunoassays, for immunoprecipitation, and for immunoblot studies. These methods will not be covered here, but are considered in Chapter 3. To our knowledge no antiserum raised against one of the lipases has been found to cross-react with the other lipase. Thus, the homologous regions are probably not strong antigenic determinants.

In relation to lipoprotein metabolism it is the catalytic activity of the lipases which is the most relevant parameter, and measurement of activity will be the topic of the rest of this chapter. *In vivo* the enzymes are immobilized and act on aggregated, heterogenous substrates under constraints that we know little of (e.g. substrate availability, product assimilation etc.). Therefore the activities recorded *in vitro* under optimized conditions may not always reflect the *in vivo* situation.

Several reviews on methodological aspects of LPL (9–11) and HL (2) are available.

2. Assay of lipase activity

2.1 Preparation of triglyceride emulsions

The substrates commonly used for measurement of LPL and HL are long-chain triglycerides emulsified with phospholipids or other surface active components (e.g. gum arabic). For LPL, the reaction rates with synthetic emulsions are comparable to those obtained with VLDL or chylomicrons. For convenience, and to increase assay sensitivity, labelled triglycerides are often used. We use [^{3}H]oleic acid labelled triolein which can be obtained from several of the radiochemical companies. ^{3}H-labelled triolein is cheaper than ^{14}C-labelled. Triolein is preferable to saturated trigylcerides (e.g tripalmitin) since it is melted at assay temperatures, and is preferable to more unsaturated triglycerides (e.g. trilinolein) because it is less prone to oxidative damage.

2.1.1 Purification of triolein

Labelled triolein usually needs to be repurified to remove traces of labelled fatty acids which otherwise cause high blank values, and to remove partial glycerides. For measurement of low lipase activities, e.g. in plasma, a low and stable blank value is crucial. For measurements of high activities, e.g. in adipose tissue extracts, a higher blank is tolerable. Partial glycerides (monoglycerides, diglycerides) are more polar than triglycerides and there-fore localize mainly at the surface of the emulsion particles, where they are more readily hydrolysed than the bulk of triglycerides. Thus, presence of partial glycerides in the labelled triglyceride preparation can lead to overestimation of the activity. On the other hand, presence of disproportion-ate amounts of unlabelled partial glycerides can suppress hydrolysis of labelled triglycerides and lead to underestimation of true lipase activity.

Methods to purify triolein by chromatography on silica gel columns (12) or by liquid–liquid partitioning (10) have been described. We use TLC on silica gel coated plates which separates the triolein from fatty acids and partial glycerides, as described in *Protocol 1*.

Protocol 1. Purification of ^{3}H-labelled triolein by TLC

1. Prepare (or buy) 0.50 mm silica gel TLC plates. Dry them at 60 °C overnight.

2. Apply the labelled triolein as a band across most of the plate (< 1 mg/cm, total amount 2.5 mCi) and unlabelled triolein, preferably partially hydrolysed, as standard in lanes on both sides (10 µg).

3. Develop the plate in a single phase system containing heptane:diethyl ether:methanol:acetic acid (85:30:3:2, v:v:v:v).

4. Cover the labelled lane with a glass plate and visualize the standard lanes by iodine vapour.

5. With a needle, mark out the area corresponding to the triglycerides (fastest migrating main band) and scrape this silica into a small glass column. Elute the triolein with 30 ml anhydrous diethyl ether. Evaporate under N$_2$ at room temperature and redissolve the triolein at a suitable concentration in heptane or mix with unlabelled triolein in heptane to a

Protocol 1. *Continued*

convenient stock concentration. Store in the freezer under N_2 in tightly sealed tubes or ampoules. The material can be used for about two months.

2.1.2 Preparation of gum arabic-stabilized labelled emulsions

Emulsions stabilized by gum arabic are recommended for assay of HL, and can also be used for measurement of LPL (see *Protocol 2*). This emulsion is prone to adsorption of non-specific proteins (e.g. from tissue homogenates) which may inhibit lipase activity by covering the oil–water interface. With LPL, the basal activity (in the absence of apolipoprotein CII or serum) is high. The stimulation by apo CII is consequently less (1.5 to 2.5-fold) than with other types of emulsions.

Protocol 2. Gum arabic-stabilized emulsion

1. Mix 25 mg unlabelled triolein (e.g. from a heptane solution containing 25 mg/ml) with ^{3}H-labelled triolein ($\sim 50 \times 10^6$ d.p.m.) in a rounded-bottom 30 mm diameter glass vessel suitable for sonication. Evaporate solvent under N_2 at room temperature.

2. Add 1 ml 10% (w/v) gum arabic in water (Sigma). Gum arabic takes some time to dissolve with magnetic stirring. The stock gum arabic solution can be stored frozen in suitable aliquots.

3. Add 1.25 ml 1 M Tris–HCl buffer pH 8.5 and 2 ml water.

4. Chill the vessel in ice-water and sonicate for 10 min in a 50% pulsed mode (MSE Soniprep 150), with a 9.5 mm diameter flat tip probe at medium setting placed a few mm below the surface of the liquid. It is important to standardize conditions of vessel geometry, volume, tip placement, energy, and time.

5. Inspect the emulsion. If oil droplets are floating on the surface, sonication was not sufficient. Try other settings.

6. Mix emulsion (6.7 mM triolein) with other constituents of the incubation medium. Use this emulsion on the same day. Similar results are usually obtained after storage overnight, but this is not recommended.

2.1.3 Preparation of phospholipid-stabilized labelled emulsions

Zilversmit *et al.* (13) pointed to the possibility of dispersing triglycerides in glycerol with phospholipids as stabilizers. Nilsson-Ehle and Schotz (12) used this to develop a stable stock emulsion for assay of LPL (see *Protocol 3*).

Protocol 3. Phospholipid-stabilized emulsion in glycerol

1. In a wide glass text tube mix 600 mg unlabelled triolein with ^{3}H-labelled triolein ($\sim 5 \times 10^9$ d.p.m.) and 36 mg phosphatidyl choline (egg yolk, Sigma). Evaporate solvents under N_2.
2. Add 10 ml (12.5 g) glycerol.
3. Chill the tube in ice water. Homogenize continuously with a Polytron homogenizer (Kinematica) for 5 min. The emulsion should be optically clear and can be stored for at least 6 weeks at room temperature. The theoretical concentration of triolein in the stock emulsion is 68.7 mM.
4. Substrate solutions for assay are prepared daily from the stock emulsion.

2.1.4 Labelling of Intralipid with [^{3}H]triolein

In our hands, the most reproducible measurements of LPL are obtained with the commercial emulsion Intralipid® and manual titration of the released fatty acids. The sensitivity in such assays is, however, too low for many applications. This can be overcome by incorporation of a labelled triglyceride into Intralipid by sonication as described in *Protocol 4*. The activities recorded from release of labelled fatty acids are comparable to those recorded by titration. This indicates that the labelled triolein mixes well with the Intralipid lipids. The sonicated emulsion is, however, stable for only about two weeks. Intralipid 10% is an emulsion of soya bean triglycerides in egg yolk phosphatidyl choline used for parenteral nutrition. It contains 100 g triglycerides and 12 g phospholipid per litre. Presumably, other lipid emulsions for parenteral nutrition can be used.

Protocol 4. Procedure for the labelling of Intralipid with ^{3}H-labelled triolein

1. Evaporate labelled triolein ($\sim 10^9$ d.p.m.) under N_2 on the walls of a round-bottom 30 mm diameter glass vessel suitable for sonication.
2. Add 5 ml Intralipid (10%, Kabi).
3. Chill the vessel in ice-water and sonicate for 10 min in a 50% pulsed mode (MSE Soniprep 150 at medium setting with a flat tip 9.5 mm probe, placed a few mm below the surface of the liquid). It is important to standardize conditions of vessel geometry, volume, tip placement, energy, and time.
4. Inspect the emulsion. It usually does not change appearance from normal Intralipid. After sonication the emulsion is less stable, but can be stored at 4 °C for 1–2 weeks with unchanged substrate properties. The triglyceride concentration of the emulsion is approximately 115 mM.

2.2 Assay conditions

The reactions involved in lipase assays are complex. Triglycerides are sequentially hydrolysed to diglycerides, monoglycerides, and free glycerol. The action on each of these intermediary products differs, both because of properties of the enzymes' active site (low activity against 2-monoglycerides, but high activity against 1,3-monoglycerides) and because of the differing physical properties of the products. Usually the rate-limiting step is hydrolysis of the first ester bond in the triglyceride (14).

Below we discuss briefly some properties of the lipases relevant to measurement of their activities. References to original work on which this is based can be found in general reviews (1, 2, 14).

2.2.1 Substrate concentration

The relevant parameter is the amount of lipid–water interface, not the total amount of emulsified lipid. One cannot use data on substrate requirements from the literature, but must determine this experimentally, because it depends on the physical properties of the home-made emulsion. Furthermore, the substrate dependence differs for LPL from different sources; e.g. human LPL requires higher substrate concentration than bovine LPL. Enzyme sources containing lipid-binding proteins may require higher substrate concentrations, because the proteins compete with the lipase for the interface. In practice, 2–4 mg triglyceride per ml will often provide saturating or close to saturating amounts of substrate for assay of LPL. HL often requires more. One should not use a large excess of substrate emulsion, because this will increase the blank.

2.2.2 Albumin

Both LPL and HL are inhibited by fatty acids. It is therefore necessary to include albumin to bind the fatty acids. Furthermore, albumin at concentrations of a few mg per ml is useful to prevent adsorption of the lipases to glassware. The molar ratio of fatty acids to albumin should not exceed 5:1 at any time. Incubation systems for triglyceride hydrolysis usually contain 30–60 mg albumin per ml. For most purposes it is not necessary to use fatty acid-free albumin.

Some commercial albumin preparations contain traces of lipids and apolipoproteins (15). This can have profound effects on the kinetics of the lipolytic reactions, and on the product profiles. We use bovine albumin, fraction V powder (Sigma). For each new batch we check that the albumin, in high concentrations, does not cause inhibition of any of the lipases.

2.2.3 pH

The observed pH dependency for LPL activity often follows a bell-shaped curve with optimum around pH 8.5. The shape of the curve varies, however, considerably with the assay conditions. Closer studies have revealed that the rate of the active site reaction itself increases with pH at least to pH 10 (14). The decrease of observed activity above pH 9 can be explained by decreased binding of LPL to the lipid–water interface, and/or decreased stability of the enzyme. No detailed study of the pH-dependency of HL has been done but it is assumed that it is similar to that of LPL.

The activity of both lipases is usually measured at alkaline pH (8.2–8.6) because the higher rate makes the assays more sensitive. It should be pointed out, however, that the enzymes have high activity also at physiological pH (7.4), about half of that at pH 8.5. Other aspects of the reaction may be sensitive to pH, such as lipid packing, lipid–protein interactions, and the rate at which partial glycerides/lysophospholipids are isomerized. Therefore, pH 7.4 is recommended for studies which explore physiological reactions, such as lipoprotein interconversions.

2.2.4 Temperature

In an early study (16) it was noted that LPL-catalysed release of fatty acids was linear with time at room temperature, but not at 37 °C. This was because the enzyme was not stable at 37 °C under the assay conditions used. This important observation has been overlooked in many subsequent studies. Unless there is a specific reason to use 37 °C, such as for studying lipid transitions, it is recommended that incubations be carried out at 25 °C.

2.2.5 Buffer composition

It is considered characteristic for LPL ('the salt-sensitive lipase') that it is inhibited by 0.5–1 M NaCl. It is important to realize that this is due to an irreversible denaturation of the enzyme molecule. If this is avoided, for instance by running the incubation at low temperature, the enzyme can exert full catalytic activity even in the presence of 1 M NaCl. For most purposes NaCl concentrations of 0.05–0.15 M result in optimal activity and stability of LPL for kinetic studies. The activity of HL ('the salt-resistant lipase') is not affected much by 1.0 M NaCl.

2.2.6 Activator

With triglyceride emulsions as substrates the activity of LPL (but not the activity of HL) is increased several fold by apolipoprotein (apo) CII. Detailed protocols for preparation of apo CII can be found in Jackson and Holdsworth (17). The concentration of apo CII required varies, but 1 µg/ml (10^{-7} M) is usually enough for maximal or near-maximal activation. The relevant parameter is probably the two-dimensional concentration of apo CII at the lipid–water interface. Hence, more apo CII is needed with higher concentrations of lipid substrate. Apo CII tends to aggregate in solution. To prevent this we use 3 M guanidinium chloride in 10 mM Tris–HCl, pH 8.2 in stock solutions (4.4 mg/ml = 0.5 mM) of the apolipoprotein. The effect of apo CII varies considerably with the substrate system. With most triglyceride emulsions the enzyme displays activity even without the activator. This basal activity is usually 5–20%, but can approach 100% of the maximal activity.

For assay purposes, whole serum or HDL can be used to provide activator. Similar results are obtained with plasma as with serum. Due to the presence of low lipase activities, the serum should be heat inactivated at 56 °C, 30 min. Serum from most mammals, with the exception of guinea pigs, can probably be used but the most common sources are fasted humans and starved rats. Due to individual variation in the amounts of activator it is recommended to prepare a large pool, dialyse it against physiological saline, and test out the optimal amount to be used in the assay (usually 5–10% v/v). The serum can be stored at −20 °C.

In contrast, HL is inhibited by apo CII and other lipid binding proteins in serum, which transfer to the emulsion particles and cover their surface. Furthermore, HL has a preference for HDL particles, which hence act as a competing substrate.

2.2.7 Other proteins

It has been reported that the activities of LPL and HL are inhibited by a variety of proteins. The mechanism(s) of the inhibition has not been explored in detail. It is probably non-specific since inhibition is seen with many different proteins. In no case has a specific protein–protein interaction between the lipase and an inhibitory protein been demonstrated. The common denominator seems to be that the proteins bind to the surface of the substrate droplets.

A corollary is that proteins which are added to the assay together with the lipases, for example in tissue homogenates or plasma, can inhibit the reaction. Therefore one should use a sensitive assay so that only small volumes of sample are needed.

Whenever the assay is applied to a new enzyme source, it is imperative to test that lipase activity is linearly related to the amount of enzyme. Unfortunately, this is sometimes neglected. The reader is referred to a recent paper by Vannier *et al.* (18), who demonstrate and discuss the effects of this type of error on results and conclusions of earlier work.

2.2.8 Heparin

The LPL–heparin complex is more stable and soluble than the enzyme alone, and is catalytically active (4). Therefore heparin is included in most buffers used for extraction of LPL from tissue sources, and for incubation of the enzyme in kinetic studies. Dilute solutions of LPL can usually be stabilized by heparin at concentrations of 1 μg/ml (corresponding to ∼ 0.1 μM, and ∼ 0.15 IU/ml for most pharmaceutical preparations).

There are reports that heparin may change the kinetic parameters for LPL and HL under some conditions, probably by competing with the substrate emulsion for binding of the enzymes. In the usual assay for HL with a high concentration of NaCl, the interaction between HL and heparin will be disrupted, and the presence of heparin will be of no importance.

2.2.9 Selective measurement of LPL and HL

There are two approaches to this, either to use assay conditions which favour one of the enzymes or to use antibodies that inhibit the activity of one enzyme. The activity of LPL can readily be suppressed by 1 M NaCl, and the activity recorded (with the gum arabic-stabilized emulsion in the absence of serum or apo CII) is a good measure of HL. This is the method we advocate. Note that the mechanism of suppression is an irreversible conformational change in LPL. For this to work the pH should be relatively high (8.5–8.6), and the temperature should be at least 25 °C. An alternative approach has been to add protamine. This denies LPL the stabilization afforded by heparin (which is consumed by the protamine). Again the pH and temperature must be relatively high for this to work.

Nilsson-Ehle and Ekman (19) described a method for selective measurement of LPL and HL based on specific conditions for preparation of the substrate emulsions. Another approach is to inhibit HL by sodium dodecyl sulphate (20). Separation of the two lipases on heparin–agarose before assay has also been used (21). In our hands these methods have not been reliable, but can result in incomplete and varying recovery of lipase activity, depending on the composition of the samples used. We therefore advocate immuno-inhibition of HL for routine assays of LPL. This is accomplished by mixing the sample with an appropriate amount of antiserum or preferably immune IgG and incubating for 2 h on ice. LPL in plasma or in tissue extracts in the heparin–detergent buffer (see below) does not lose appreciable activity during this time. Centrifugation before assaying is not needed and does not change the results. As a control, incubate some samples after immuno-inhibition under the conditions used for HL-determination. If all is well there should be little or no activity. Otherwise one has to determine whether the immunoinhibition was not complete (> 95%), or the LPL was not fully suppressed by 1 M NaCl.

To prepare HL for immunization, the best sources are post-heparin plasma or heparin-perfusates of livers. Detailed protocols can be found in Ehnholm and Kussi (2). If the antiserum is to be used only for the selective assay, the antigen does not have to be homogenous. The important consideration is that there is no LPL. To ensure this, we treat the preparation with antiserum to LPL (see below) before the final purification step. Immunization is done by standard techniques (see Chapter 3). Rabbits respond well to both human and rat HL.

Sometimes there is a need to differentiate LPL from other lipases, for instance in tissue homogenates. Some information can be obtained by testing the effects of including 1 M NaCl in the assay, and/or excluding serum (apo CII). This should suppress the activity if it is due to LPL. A more direct approach is to pretreat the enzyme source with antibodies to LPL. A useful manoeuver has been to use bovine LPL to immunize chickens (22). This has given us crossreacting antisera, which inhibit human, rat, mouse, rabbit, and hamster LPL.

Convenient assay systems for LPL and HL are given in *Protocols 5* and *6*, respectively.

Protocol 5. Assay for LPL

1. Prepare assay medium: 0.3M Tris with 0.2 M NaCl, 0.02% (w/v) heparin and 12% (w/v) bovine serum albumin. Adjust the pH to 8.5 with HCl. This solution can be stored at −20 °C.

2. Mix appropriate volumes of assay medium, heat-inactivated serum and labelled emulsion at room temperature. This solution is stable for one day. The amount of serum should be tested for each batch of serum used. A typical composition is assay medium: rat serum: ^{3}H-labelled Intralipid 10:1:1. Pipette 120 µl into each tube for assay.

3. The total volume should be 200 µl. Hence, a maximum of 80 µl sample can be added. For plasma and tissue homogenates one should preferably use less than 10 µl to avoid interference by other components in the sample. The rest of the volume can be made up with water.

4. Let the tubes incubate at 25 °C for 5 min before addition of the enzyme source. Continue incubation in a shaking water bath. For longer times (1–2 h) it may be necessary to cover the tubes to prevent evaporation.

5. Run blank incubations (without lipase) for each assay, and preferably also reference sample(s). Assay samples in duplicate or triplicate.

6. Stop the reaction after the desired time by addition of solvents for one of the extraction procedures.

Protocol 6. Assay for HL

1. Mix one batch of the gum arabic-stabilized emulsion with 2.5 ml each of 5 M NaCl and of 10% (w/v) bovine serum albumin in water (titrated to pH 8 with NaOH, stored frozen). Add 3.25 ml water. Total volume will be 12.5 ml. Use on the same day only.

2. Pipette 150 µl substrate mix into each tube. Add water and/or buffer to make 200 µl when the sample has been added.

3. Incubate at 25 °C for 5 min before the enzyme source is added.

4. Continue as under assay for LPL (*Protocol 5*, steps 5 and 6).

2.3 Extraction and quantitation of fatty acids

Unlabelled fatty acids can be determined by manual or automatic titration, by enzymatic methods (e.g. NEFA-QUICK, Boehringer Mannheim) or by nickel-binding assays (23).

2.3.1 Extraction of product fatty acids by the method of Belfrage and Vaughan (24)

To extract and measure fatty acids released in assays with labelled substrate lipids the method of Belfrage and Vaughan is suitable for most routine purposes because it is simple and comparatively rapid. Lysophospholipids are partially extracted into the upper phase. This method can therefore be used for studies of phospholipase activity with glycerol- or choline/ethanolamine-labelled phospholipids as substrate.

Protocol 7. Procedure for the extraction of product fatty acids (24)

1. Incubate in glass tubes. For a 0.2 ml assay system 13 × 100 mm tubes are suitable. It is difficult to clean used tubes properly. Therefore disposables are recommended, at least for assay of samples with low activity.

2. Stop the reaction by addition of 3.25 ml methanol:chloroform:heptane (1.41:1.25:100. v/v/v). Use a dispenser flask.

3. Add 1 ml of 0.1 M sodium carbonate buffer pH 10.5 (from a dispenser). Mix vigorously on a Vortex-type mixer for a couple of seconds.

4. Centrifuge the tubes (preferably in a swing-out rotor at 1500g for 10 min) to separate the upper alkaline/water phase containing the fatty acids from the lower chloroform heptane-phase which contains triglycerides and partial glycerides.

5. Transfer a sample (0.4–1 ml) from the upper phase to a scintillation-counting vial and add an adequate amount of scintillation cocktail (usually 5–10 volumes). Mix. For some cocktails it may be necessary to add acetic acid to neutralize the alkaline sample. Otherwise there will be marked chemiluminescence.

6. Determine extraction efficiency using a trace amount of labelled fatty acid added to assay mix. The extraction factor (usually around 50%) should be checked occasionally.

7. To determine specific radioactivity of the substrate, count four aliquots of the incubation mixture. Add upper phase (from a blank incubation) to get the same conditions for counting as for the samples. Calculate specific radioactivity as d.p.m./nmol fatty acid. Triglyceride concentration is calculated from the known composition of the medium, or determined by an enzymatic kit.

2.3.2 Extraction of lipids by the method of Dole (25) followed by separation of fatty acids into alkaline ethanol

For assay of low lipase activities, when a low blank is required, we find that the following method gives more reproducible data. It is based on extraction of total lipids according to Dole (25) followed by separation of the fatty acids into an alkaline ethanol phase which is repetitively washed with heptane to remove contaminating glycerides. This gives essentially complete extraction of fatty acids and usually a 10-fold lower blank than the Belfrage–Vaughan method.

Protocol 8. Procedure for the extraction of lipids according to Dole (25) followed by separation of fatty acids into alkaline ethanol

1. Same as in *Protocol 7*.
2. Add 1.25 ml of Dole's solution (chloroform:heptane:0.5 M H_2SO_4, 40:10:1) with a dispenser.
3. Then add 0.75 ml heptane and 0.55 ml water. Mix vigorously.
4. Centrifuge to separate the phases. Transfer 0.8 ml of the upper phase (containing total lipids) to new tubes with 1 ml alkaline ethanol (ethanol 95%:water:2 M NaOH, 500:475:25). Add 3 ml hexane. Mix.
5. Centrifuge 1500*g* for 2 min to separate phases. Discard the upper heptane phase (which contains the unhydrolysed triglycerides). This is most conveniently done by coupling a glass pipette to a water-operated vacuum source. Collect the heptane in an intervening vacuum flask.
6. Add 3 ml heptane again and repeat step 5. Discard all heptane.
7. Take a suitable sample (0.4–0.8 ml) of the remaining alkaline ethanol phase (which should contain only fatty acids) to a scintillation vial with an appropriate amount of scintillation cocktail (5–10 volumes).
8. Determine extraction efficiency and substrate specific radioactivity as in *Protocol 7*.

---

3. Preparation and storage of samples

3.1 Plasma

Injection of heparin releases LPL and HL from their tissue sites by forming soluble enzyme–heparin complexes (4). For HL, it appears that an equilibrium is established between HL at binding sites in the liver and HL circulating with heparin in plasma. In contrast, LPL is rapidly turned over by uptake in the liver. Therefore, the LPL activity represents a balance between release to and uptake from blood. The release is promoted and the uptake is impeded, but not abolished, by heparin. HL activity rises rapidly, LPL activity more gradually after injection of heparin. At longer times the activities decline mainly due to clearance of heparin. For most mammals a dose of 50–100 IU heparin/kg body weight will give maximal lipase release. Samples are usually taken after 10–15 min, when both lipases display maximal or close to maximal activity. The procedure for collecting post-heparin plasma is given in *Protocol 9*.

There are rather large differences between individuals in LPL and HL activity in post-heparin plasma. Measurement in the same individual on different occasions usually gives similar values.

Protocol 9. Collection of human post-heparin plasma

1. Get ethical approval.

2. Inject 100 IU heparin/kg body weight intravenously.

3. After 15 min, take blood in heparinized vacutainer tubes.

4. Chill immediately in ice-water to preserve lipase activities.

5. Pellet the blood cells by low-speed centrifugation. Collect the plasma. If the plasma is hypertriglyceridaemic, a fat cake may form at the surface after centrifugation. Some lipase will bind to this lipid. Therefore, recover it together with the plasma for proper assay of lipase activities.

6. Measure lipase activities immediately or after storage at −70 °C. A general experience is that the activities are rather stable, however, repeated freezing and thawing are not recommended.

The activities of LPL and HL in basal plasma are low. Collect and handle samples as in *Protocol 9*, steps 3–6. Heparin should be used as anticoagulant rather than citrate or EDTA because it stabilizes LPL (and possibly also HL).

3.2 Tissue extracts

The two main approaches to estimate tissue LPL activity are to measure total activity and/or to measure the heparin-releasable activity. It is assumed that heparin releases LPL from the vascular bed, and that this is the functional enzyme which participates directly in lipoprotein metabolism. The rest is probably intracellular. For instance in perfused rat hearts heparin rapidly releases 5–25% of total LPL activity, depending on the nutritional condition (26).

3.2.1 Heparin eluates of tissue pieces

Several protocols for elution of LPL from tissue pieces have been described (8, 27–29). They have been adapted to small specimens obtained by needle biopsies. The activity recovered is assumed to be the sum of endothelial LPL and some intracellular LPL which has been secreted during the extraction, balanced by inactivation/degradation in the medium. Iverius and Östlund-Lindqvist (10) recommended extraction of adipose tissue pieces (50 mg) in six volumes 25% (v/v) human serum in Kreb's–Ringer phosphate buffer with 50 μg (about 7.5 IU) heparin per ml. After 30 min at 37 °C, 0.2 ml samples were taken for assay of enzyme activity. This method takes advantage of the stabilizing effects of both heparin and serum components (probably lipoproteins) on LPL. Iverius and Brunzell (30) recommended extraction at 4 °C to minimize inactivation. Lithell and Boberg (27) use a buffer with high glycine content to stabilize LPL and include the labelled triglyceride substrate so that extraction and hydrolysis occur concomitantly.

3.2.2 Total tissue activity

In earlier studies LPL activity was often measured in extracts of acetone:ether powders of tissues (31, 32). This method had several advantages:

- LPL activity is rather stable in the powder, if stored (desiccated) at −70 °C.
- Lipids are removed which may otherwise interfere with the assay.
- Some other lipases, e.g. hormone-sensitive lipase in adipose tissue, are inactivated.

The disadvantage is that recovery of LPL activity on extraction of the powders is usually not complete. We recommend the same buffer as for tissue homogenates, see below. Let the powder soak in the buffer before it is dispersed by homogenization or sonication. Centrifuge to pellet insoluble material and assay lipase activity in the clear supernatants.

LPL has also been assayed in tissue homogenates prepared in buffers without heparin and/or detergents. This is not recommended, because significant amounts of lipase activity remain insoluble.

The most efficient extraction/solubilization of LPL from homogenates is obtained with buffers containing detergents (10, 11, 18, 29, 33). A further advantage is that LPL is usually stable for days at 4 °C, and for hours at room temperature in these buffers. Several different detergent cocktails have been described. Most also contain protease inhibitors. The one described in *Protocol 10* was designed to be suitable also for immunoassay and immunoprecipitation. This detergent containing buffer can also be used for solubilization of cultured cells. It is excellent for dilution of purified LPL, e.g. for assay purposes. LPL solubilized as described above can bind to heparin–agarose but for this purpose heparin should be omitted from the buffer. Whenever the protein concentrations of the extracts are low, increased recovery and stability is obtained if bovine serum albumin (1 mg/ml) is included in the buffer.

The heparin–detergent buffer works well also for extraction of HL from liver and steroid-producing glands.

Protocol 10. Homogenization of tissues in detergent-containing buffer

1. Prepare buffer: 25 mM ammonia adjusted with HCl to pH 8.2 containing 5 mM EDTA and, per ml, 10 mg Triton X-100, 1 mg SDS, 5 IU heparin, 10 µg leupeptin (Boehringer), 1 µg pepstatin, and 25 KIU Trasylol (Sigma). Stock solutions of leupeptin and pepstatin (1 mg/ml) are made in ethanol (95%). The proportion between Triton and SDS (w/w) should be at least 10. Higher proportions of SDS will denature LPL.
2. Homogenize tissues in cold (4 °C) buffer with a Polytron (Kinematica) homogenizer (1 g of tissue, 9 ml of buffer). Keep the homogenates on ice.
3. Centrifuge (about 20 000g for 20 min at 4 °C). Recover the clear supernatant. With adipose tissue, recover the clear phase between the floating lipid and the pelleted material (use a syringe).
4. Assay for LPL activity within 6 h. The extracts can usually be frozen but the recovery has to be checked for each experimental situation.

4. Standardization of assay and quality control

Different assay procedures result in widely varying values for lipase activities. For instance, Taskinen (7) made a survey of the literature and found that values reported for LPL in post-heparin plasma from normal individuals differed about 10-fold between laboratories. The difference for adipose tissue LPL was even greater (6–110 mU/g).

Using the procedures we describe above, the intra-assay variation, i.e. the reproducibility for repetitive measurements of a sample in the same assay, is usually low (± 5%). The variation between assays can, however, be substantial, due to the complex nature of the substrate, up to ± 25%. It is important to quantify the specific radioactivity of the substrate accurately (d.p.m. nmol fatty acid). As a quality control one should occasionally determine the amount of fatty acids released in some samples by manual titration, and compare this to the release of fatty acids calculated from radioactivity. It is advisable to include a reference sample in each assay and use this to correct for interassay variation. Whenever the reference deviates by more than 25% from the expected value one should consider whether something was not optimal during the assay. Due to the instability of the enzymes, it is hard to find a good standard. For LPL we and others (10) use bovine skim milk frozen in portions at −70 °C. For HL, and for assays of LPL in human pre- or post-heparin plasma, a batch of human post-heparin plasma stored at −70 °C is the best reference (8).

Unfortunately, several different units are used to express lipase activities.

We advocate definition of a mU as the release of 1 nmol fatty acid per min at 25 °C and pH 8.5. This is becoming the most frequently used unit.

Our experience is that the assay continuously has to be kept under thorough control since there are many components of the system which may vary.

4.1 Check-list for lipase assay

(a) Is the amount of substrate saturating, or at least close to saturating, under the conditions used? This can differ between tissues, and between species.

(b) What is the degree of substrate consumption? Adjust incubation time and sample volume so that hydrolysis does not exceed 10% for any sample.

(c) Is the amount of albumin sufficient to bind all the liberated fatty acids even at maximal hydrolysis?

(d) For assay of LPL, is the amount of serum/apo CII optimal? All sera stimulate when added in low amounts, but with some sera the activity decreases when higher amounts are added. Avoid such sera.

(e) Is the assay linear with time? If not, change to a temperature and pH where the enzyme is stable.

(f) Is the assay linear with the amount of enzyme? If not, check whether any component in the enzyme preparation (e.g. tissue proteins, detergents) causes inhibition. At what concentration? Use small aliquots of the enzyme preparation to avoid this source of error.

(g) Is the enzyme stable over min, hours, or days?

Acknowledgements

Our studies on the lipases are supported by the Swedish Medical Research Council (grant 13X-727). Due to restrictions of space the number of references is severely limited. We apologize to those colleagues whose important contributions are cited only indirectly.

References

1. Eckel, R. H. (1989). N. Engl. J. Med., **320**, 1060.
2. Ehnholm, C. and Kuusi, T. (1986). In *Methods in Enzymology* (ed. J. P. Segrest and J. J. Albers), Vol. 129, p. 716. Academic Press Inc., London.
3. Olivecrona, T. and Bengtsson-Olivecrona, G. (1990). *Curr. Opin. Lipidol.*, **1**, 222.
4. Olivecrona, T. and Bengtsson-Olivecrona, G. (1989). In *Heparin* (ed. D. A. Lane and U. Lindahl), p. 335. Edward Arnold, London.
5. Kirchgessner, T. G., Chaut, J. C., Heinzmann, C., Etienne, J., Guilhot, S.,

Svenson, K., Ameis, D., Pilon, C., D'Auriol, L., Andalibi, A., Schotz, M. C., Galibert, F., and Lusis, A. J. (1989) *Proc. Natl. Acad. Sci. USA*, **86**, 9647.

6. Ekström, B., Nilsson, Å., and Åkesson, B. (1989). *Europ. J. Clin. Invest.*, **19**, 259.
7. Taskinen, M. R. (1987). In *Lipoprotein Lipase* (ed. J. Borensztajn), p. 201. Evener, Chicago.
8. Kuusi, T., Ehnholm, C., Viikari, J., Härkönen, R., Vartainen, E., Puska, P., and Taskinen, M.-R. (1989). *J. Lipid Res.*, **30**, 1117.
9. Nilsson-Ehle, P. (1987). In *Lipoprotein Lipase* (ed. J. Borensztajn), p. 59. Evener, Chicago.
10. Iverius, P. H. and Östlund-Lindqvist, A. M. (1986). In *Methods in Enzymology* (ed. J. P. Segret and J. J. Albers), Vol. 129, p. 691. Academic Press Inc., London.
11. Bengtsson-Olivecrona, G. and Olivecrona, T. (1991). In *Methods in Enzymology* (ed. E. A. Dennis), Vol. 197, p. 345. Academic Press Inc., London.
12. Nilsson-Ehle, P. and Schotz, M. C. (1976). *J. Lipid Res.*, **17**, 536.
13. Zilversmit, D. B., Salky, N. K., Trumbell, M. L., and McCindless, E. L. (1956). *J. Lab. Clin. Med.*, **48**, 386.
14. Olivecrona, T. and Bengtsson-Olivecrona, G. (1987). In *Lipoprotein Lipase* (ed. J. Borensztajn), p. 15. Evener, Chicago.
15. Deckelbaum, R. J., Olivecrona, T., and Fainaru, M. (1980). *J. Lipid Res.*, **21**, 425.
16. Greten, H., Levy, R. I., and Fredrickson, D. S. (1968). *Biochim. Biophys. Acta*, **164**, 185.
17. Jackson, R. L. and Holdsworth, G. (1986). In *Methods in Enzymology* (ed. J. P. Segret and J. J. Albers), Vol. 128, p. 288. Academic Press Inc., London.
18. Vannier, C., Deslex, S., Pradines-Figuères, A., and Ailhaud, G. (1989). *J. Biol. Chem.*, **264**, 13199.
19. Nilsson-Ehle, P. and Ekman, R. (1977). *Artery*, **3**, 194.
20. Baginsky, M. L. and Brown, W. V. (1977). *J. Lipid Res.*, **18**, 423.
21. Boberg, J., Augustin, M. L., Tejada, P., and Brown, W. V. (1977). *J. Lipid Res.*, **18**, 544.
22. Olivecrona, T. and Bengtsson-Olivecrona, G. (1983). *Biochim. Biophys. Acta*, **752**, 38.
23. Huang, J., Roheim, P. S., Sloop, C. H., and Wong, L. (1989). *Analyt. Biochem.*, **179**, 413.
24. Belfrage, P. and Vaughan, M. (1969). *J. Lipid Res.*, **10**, 341.
25. Dole, V. P. (1956). *J. Clin. Invest.*, **35**, 150.
26. Borensztajn, J. (1987). In *Lipoprotein Lipase* (ed. J. Borensztajn), p. 133. Evener, Chicago.
27. Lithell, H. and Boberg, J. (1970). *Biochim. Biophys. Acta*, **528**, 58.
28. Taskinen, M. R., Nikkilä, E. A., Huttunen, J. K., and Hilden, H. (1980). *Clin. Chim. Acta*, **104**, 107.
29. Kern, P. A., Marshall, S., and Eckel, R. H. (1985). *J. Clin. Invest.*, **75**, 199.
30. Iverius, P. H. and Brunzell, J. D. (1985). *Am. J. Physiol.*, **249**, E107.
31. Robinson, D. S. (1963). *Adv. Lipid Res.*, **1**, 133.
32. Stewart, J. A. and Schotz, M. C. (1974). *J. Biol. Chem.*, **249**, 904.
33. Semb, H. and Olivecrona, T. (1986). *Biochim. Biophys. Acta*, **921**, 104.

7b

Cholesterol esterifying enzymes—lecithin:cholesterol acyltransferase (LCAT) and acylcoenzyme A: cholesterol acyltransferase (ACAT)

MICHAEL P. T. GILLETT and JAMES S. OWEN

1. Introduction

Unesterified cholesterol (UC) is an important structural component of cell membranes, modifying their fluidity, and of the surfaces of lipoprotein particles; it is also the precursor for steroid hormones and bile acids. The apolar nature of cholesteryl ester (CE) molecules largely excludes them from lipoprotein surfaces and from cell membranes and they form lipid droplets, either within the core region of lipoproteins or intracellularly. The CE of lipoproteins can be regarded as a transportable form of UC and that of cells as a storage form. Lecithin:cholesterol acyltransferase (LCAT, EC 2.3.1.43) is responsible for the formation of most lipoprotein CE and acylcoenzyme A: cholesterol acyltransferase (ACAT, EC 2.3.1.26) for intracellular CE.

2. Lecithin:cholesterol acyltransferase

The reaction catalysed by LCAT is shown in *Figure 1*. The enzyme is secreted by the liver, requires apolipoprotein (apo) AI and apo AIV for optimal activity and appears to act preferentially on the smaller high-density lipoprotein (HDL) particles, including pre-beta HDL and HDL_3. Once synthesized, CE moves into the interior of the HDL particle and some is exchanged for triglyceride (TG) in other lipoproteins by cholesteryl ester transfer protein. The transferred CE accumulates in low density lipoproteins (LDL) and cells requiring UC take these up via the LDL-receptor pathway. By its action, LCAT helps HDL to accept additional UC either from cell membranes or from other lipoproteins; this allows further esterification and eventual formation of the larger, CE-rich HDL_2 and HDL_1 particles. Because these particles deliver their CE to the liver, LCAT can be considered to have

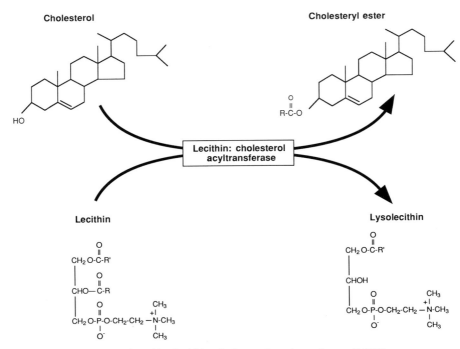

Figure 1. Reaction catalysed by lecithin:cholesterol acyltransferase (LCAT)

an anti-atherogenic action by driving centripetal cholesterol transport via this HDL carrier system.

Measurement of LCAT is of interest, not only because it plays a central role in lipoprotein metabolism, but also because several clinical conditions, such as liver and renal disease and familial LCAT deficiency, are associated with low or even absent levels of plasma LCAT. The resulting profound lipoprotein abnormalities have several adverse effects on cell membranes and cellular function. It is reported that LCAT measurement is more sensitive than conventional liver function tests and is a good prognostic indicator in liver transplantation; its assay may also be useful in studies of hyper-lipidaemias. Three ways to measure LCAT are described and discussed below, but it is worth noting that in man the percentage of plasma total cholesterol as CE is remarkably constant (70–74%) and that any reduction will invariably indicate low LCAT activity.

2.1 Determination of endogenous cholesterol esterification in plasma

Early studies of cholesterol esterification in human plasma measured the decrease in UC concentration following prolonged incubation (18–24 h) at

37 °C. These assays reflect not only the plasma concentration of LCAT but also the influence of factors such as substrate depletion and product inhibition—because of difficulties in accurately quantitating UC at that time, it was not possible to measure the small decrements during the initial linear phase of esterification (< 1 h). Two methods have evolved to determine initial rates of plasma UC esterification.

2.1.1 Radiochemical method

LCAT-catalysed formation of radioactive CE following addition of tracer amounts of labelled UC to the plasma is potentially a very sensitive method to measure the initial rate of UC esterification. Unfortunately, early procedures failed to ensure adequate isotopic equilibration of labelled UC with lipoprotein UC and, therefore, did not truly measure the esterification rate of endogenous UC. This problem was resolved in 1971 by Stokke and Norum (1): in a preliminary step, LCAT in the test plasma was temporarily inhibited during the period of isotopic equilibration and then reactivated for incubation within the initial linear phase of esterification. One potentially important objection to this procedure is now recognized: during the pre-incubation step neutral lipids are transferred between lipoproteins, even in the absence of LCAT activity, and this presumably modifies their substrate qualities for LCAT. Nevertheless, the basic method is widely used and a detailed procedure, including certain modifications, is presented below. Note that the crucial step in the assay is formation of the labelled UC emulsion; poor preparation may result in incomplete equilibration and give false low values for UC esterification rates.

i. Plasma samples

Anticoagulate blood samples with solid Na_2EDTA (1 mg/ml), cool them immediately on ice and separate the plasma in a refrigerated centrifuge. As an alternative, sodium citrate (3.1 mg/ml) may be used but heparin should be avoided since it may inhibit LCAT. Serum should not be used since significant UC esterification can occur during the time taken for clot formation and retraction. Samples may be tested immediately or they can be stored at −20 °C or lower for several months without loss of activity. They should not be kept, however, at 4 °C for any length of time nor repeatedly thawed and refrozen before assay. Because minor variations in apparent UC esterification may result from the use of different labelled UC emulsions, it is best to assay plasmas from different subject groups with the same substrate preparation. If this is not possible for a given study, a standard plasma kept frozen (−70 °C) in batches should be included in each assay run to monitor substrate performance.

ii. Preparation of labelled UC/albumin emulsion and other reagents

Unless otherwise stated all reagents required for this and the following LCAT and ACAT assays may be obtained from Sigma Chemical Co., whilst all radiolabelled materials can be purchased from Amersham International Plc. For studies of human LCAT, human serum albumin (HSA) (essentially fatty acid free) is used, but for animal studies the equivalent grade of bovine serum albumin (BSA) may be substituted. [4-^{14}C]UC is preferred to [7-^{3}H]UC and its radiochemical purity, if in doubt, should be verified by TLC (Section 2.1.1.*iv*). If more than 0.1% of the radioactivity is found to run in the position of CE, then the whole batch must be repurified by preparative TLC and extracted into redistilled hexane.

Follow the procedure in *Protocol 1* to prepare the labelled UC/albumin emulsion; this standard format must be rigorously adhered to because of its critical importance to the method.

Protocol 1. Preparation of radiolabelled cholesterol–albumin emulsion[a]

1. Prepare a 5% (w/v) solution of human serum albumin (HSA) in 0.2 M sodium phosphate buffer (pH 7.4) and heat at 56°C for 30 min to destroy any endogenous LCAT activity.

2. Centrifuge the solution (500g, 15 min) and transfer the supernatant to a clean tube; weigh tube and contents.

3. For each 1 ml of HSA solution, transfer 2 μCi of [^{14}C]UC or 5 μCi of [^{3}H]UC to a clean tube and evaporate the solvent under N$_2$.

4. Redissolve the labelled UC in 100 μl of acetone, take it up into a pipette and add dropwise to the HSA solution whilst vortexing.

5. Place the tube in a 20 °C water bath and carefully direct a stream of N$_2$ onto the surface of the HSA solution to remove acetone; vortex the tube after 10 and 20 min.

6. After 30 min, when all traces of acetone should have been removed, reweigh the tube and contents to calculate the amount of water lost.

7. Add the lost volume of water dropwise whilst vortexing and pass through a 0.22 μm filter.

8. The emulsion is now ready for use. If required, flush with N$_2$ and store at 4 °C for up to 4 days before use; do not freeze.

[a] Note that micelles consisting of lysophosphatidylcholine and [^{3}H]UC (molar ratio 50:1) are recently reported to rapidly and efficiently label the lipoproteins in plasma with radioactive cholesterol (2).

---

iii. Incubation procedure and lipid extraction

For each plasma sample, label four tubes fitted with PTFE-lined screw caps (two test samples and two blanks—if more than two or three samples are to be assayed 'common' blanks can be used). Follow the procedure in *Protocol 2*.

Protocol 2. Incubation and lipid extraction procedures for measuring endogenous cholesterol esterification rate in plasma

1. Add 0.1 ml of test plasma and 20 µl of DTNB solution[a] to each tube, mix and preincubate at 37 °C in a water bath for 30 min.

2. Add 30 µl of labelled UC/HSA emulsion, mix and continue incubation at 37 °C for 4 h.

3. Start a stopwatch and, at 15 sec intervals, add 20 µl of mercaptoethanol solution[a] to test samples;[b] blanks receive 20 µl of saline. Vortex after each addition and continue the incubation.

4. After exactly 30 min, remove the first tube and add 4 ml of $CHCl_3$–MeOH (2:1, v/v) from a dispenser. Mix and close the tube. Repeat in turn with the other tubes at 15 sec intervals.[c]

5. Add 1 ml of H_2O to each tube, mix and centrifuge at 500*g* for 15 min to separate and clarify the chloroform and aqueous phases.

[a] Prepare 1.4 mM DTNB (5, 5'-dithiobis-(2-nitrobenzoic acid)) and 0.1 M 2-mercaptoethanol solutions in 0.2 M sodium phosphate buffer (pH 7.4) on the day of the assay.
[b] At this stage test samples will turn bright yellow.
[c] If convenient, the samples can then be stored overnight at 4 °C or −20 °C.

---

iv. Chromatography and radioactive counting

The next stage in the assay is to separate UC and CE and to count the radioactivity in each. In our view this is best achieved by thin-layer chromatography (TLC) rather than by high performance liquid chromatography (3) or by column chromatography, although the introduction of Bond Elut columns (Analytichem International) has simplified this latter technique (4). Commercial 20 × 20 cm plastic sheets coated with silica gel G (e.g. Merck; Schleicher & Schuell; Kodak) are the most convenient; their cost can be reduced by cutting them in half without loss of the excellent UC and CE separation.

Protocol 3. Chromatographic separation and radioactive measurement of UC and CE

1. Divide up each plate into ten tracks by scoring parallel lines 2 cm apart. Prepare the glass developing tank about 1 h before use by lining it with Whatman 3MM paper and adding 100 ml of the running solvent (hexane/diethyl ether/acetic acid, 90:20:1, by vol.), wetting the paper in the process. Close the tank tightly by placing a weight on top.

Protocol 3. *Continued*

2. After centrifuging the sample tubes (*Protocol 2*) remove the upper phase and denatured proteins from the interface by aspiration and discard. Transfer the lower chloroform phase to a 50 × 12 mm glass tube containing a few drops of MeOH (to ensure one solvent phase) using a clean Pasteur pipette for each sample (disposable polyethylene ones are convenient; note that it is not necessary to transfer quantitatively since the assay only measures the ratio of radioactivity in UC and CE).

3. Place the tubes in a heating block at 30–35 °C and evaporate the solvent almost to dryness under N_2. Using a microsyringe take up each sample and apply as a narrow streak about 1.5 cm wide and 2 cm up from the bottom of each track. After applying each sample, wash the microsyringe several times with chloroform–methanol (2:1, v/v). Place one or two plates in the tank, close the lid and develop to within 1 cm of the top.

4. Remove each plate, air-dry for 5 min and place briefly in a closed glass tank containing a few crystals of I_2. The separated lipids will quickly stain yellow/brown. Mark the positions of UC and CE by comparison with authentic standards (the R_F values are about 0.15 and 0.8, respectively) and allow the I_2 to evaporate in a fume cupboard.

5. Cut out each band with scissors, transfer to scintillation vials and add 10 ml of a toluene-based scintillation fluid (e.g. Cocktail T, BDH Chemicals Ltd). Vortex and count the radioactivity in each vial using a procedure appropriate for the isotope used.

---

v. Calculations

After counting the radioactivity in the UC and CE fractions, the percentage incorporated into CE is calculated and the blank value is subtracted. The results are expressed as the percentage of UC esterified per h. By measuring the concentration of UC in non-incubated test plasma (e.g. by the CHOD-PAP enzymatic method of Boehringer Mannheim) the nmol of UC esterified per ml plasma per h can be calculated. Typical values for normal human plasma are in the range 70–100 nmol/ml/h.

vi. Adaptation of method to study CE sub-classes formed by LCAT

The basic method decribed above can be extended in order to measure the incorporation of radioactivity into different sub-classes of CE. This might prove of interest in certain clinical situations in which the fatty acyl composition of plasma phospholipids is altered. It has also been used in comparative studies of LCAT in plasma of animals, since different species show characteristic profiles of CE fatty acids. The method already described is followed to separate the labelled CE by TLC; detection is by spraying with rhodamine 6G (0.01% (w/v) in acetone) since I_2 can attack fatty acyl double bonds. The CE band is eluted from the silica gel with chloroform–methanol (9:1, v/v). dissolved in hexane and re-chromatrographed by silver ion TLC (5) to separate CE sub-classes according to the number of double bonds in the fatty acid chains. The radioactivity in each sub-class is then counted and expressed as a percentage of total CE radioactivity.

vii. Studies with animal plasma
The original Stokke and Norum method (1) has also proved useful for studying UC esterification in plasma from animals. But the prospective user should be aware of several problems which can arise. First, the duration of the initial linear phase of UC esterification in the plasma of some species is shorter than for man; the optimal incubation period should be determined, therefore, by time course studies. Second, in some species, including many reptiles and fish (6), the LCAT enzyme is not fully inhibited by DTNB, thereby permitting some UC esterification during the pre-incubation period. Isotopic equilibration at 4 °C in the absence of DTNB has been described (7) and may prove useful for LCAT studies in such plasma. Third, in certain species, including mouse (8), mercaptoethanol itself has been shown to inhibit plasma UC esterification and so cannot be used. Finally, plasma of dog and a few other species contains a CE hydrolase. Unknown test plasmas should be tested, therefore, by incubation with labelled CE to assess whether false low values for UC esterification rate have been obtained.

2.1.2 Non-radiochemical methods
The introduction of more specific and sensitive assays for UC has now made it feasible to measure directly the initial rate of endogenous plasma UC esterification; this avoids uncertainty over both the completeness of isotopic equilibration and the alteration of LCAT substrates when exogenous labelled UC is pre-incubated with plasma (particularly if they are pathological specimens). Note that the decrease in UC is measured since UC unlike CE (and lecithin or lysolecithin) can be directly quantified and its percentage change, although less than 5% in human plasma, is invariably much greater than that of CE. The two main methods used are: gas–liquid chromatography (GLC) and enzymatic (cholesterol oxidase with colorimetric or fluorometric measurement); both can give good results, provided replicate analyses are made. But they do lack sensitivity especially if the esterification rate is very low, as in some diseases or in certain animals, and radiochemical methods are preferred for such samples. Nevertheless, the technique is of value for screening plasmas with suspected low activity, particularly if the incubation time is extended to 4 or 6 h and the fall in UC measured by a commercial colorimetric kit.

2.2 Determination of LCAT catalytic activity

Because the rate of esterification of endogenous UC reflects the quality and quantity of HDL substrate particles, in addition to LCAT concentration *per se*, plasma UC esterification has also been measured in the presence of an excess of exogenous substrate. The Glomset and Wright method (9), in which test plasma is incubated with 8-volumes of human plasma, pre-heated to destroy endogenous LCAT activity and radiolabelled with [^{14}C]UC, was first introduced but has inherent disadvantages. First, inter-laboratory standardization is impracticable; although the relative activities of test plasmas are retained using different substrates, the absolute value for a specific sample varies widely. Second, the heating step in the substrate preparation may alter the reactivity of the lipoproteins towards LCAT. Note also that the influence of endogenous lipoproteins cannot be totally excluded as they comprise about 10% of the total, albeit initially unlabelled. Accordingly, improvements have been proposed, initially the use of isolated HDL$_3$ but more recently a defined, optimal proteoliposome substrate (10). The methodology for such estimations of plasma LCAT activity is detailed in *Protocol 4*; test plasma is prepared as for the Stokke–Norum method described in Section 2.1.1.*i* but the incubation procedure and calculation differ.

Protocol 4. Heated plasma substrate method for the determination of LCAT activity

1. Prepare labelled substrate:
   - (a) Heat a suitable volume (10–50 ml) of fresh normal human plasma (preferably from several individuals) for 30 min at 56 °C.
   - (b) Cool and centrifuge at 20 000g for 20 min.
   - (c) Mix each ml of clear supernatant with 0.125 ml of radiolabelled UC/HSA emulsion (see *Protocol 1*) and incubate for 8 h at 37 °C and then overnight at 4 °C. This substrate can be kept at 4 °C for 1 week but is viable for several months at −20 °C or lower (usually batched as it should not be thawed and refrozen).

2. Incubate substrate as follows:
   - (a) Add 0.18 ml of labelled substrate to screw-cap tubes and cool in an ice bath.
   - (b) Add 20 μl of each test plasma to duplicate tubes and 20 μl of saline for the blanks.
   - (c) Mix and incubate all tubes for 4 h at 37 °C.
   - (d) Remove all tubes to the ice bath and quickly add 4 ml of CHCl$_3$-MeOH (2:1, v/v), mix and complete the lipid extraction and assay as described in *Protocols 2* and *3*.

3. The activity of LCAT is expressed as nmol UC esterified per ml of test plasma per h, never as % esterification. Calculate first the % esterification of radioactivity into CE (Section 2.1.1.*v*) by each test plasma (usually 2% or less) and subtract the blank value. Next measure the UC content of both test plasma and labelled substrate and calculate the total amount of UC (nmol) in each 10 ml of incubation mixture (equivalent to 1 ml of test plasma). Multiply this by the % of UC esterified and divide by 4 h to give LCAT activity as nmol/ml/h.

2.2.1. Proteoliposome substrate method

This chemically defined substrate consists of apo AI:lecithin:labelled UC in the molar proportions of 0.8:250:12.5 and obviates inherent problems associated with heated plasma substrates. The method has allowed the assay of LCAT activity to be optimized and standardized for plasma. It is also suitable for assaying LCAT during a purification procedure or, after concentration, in media in which hepatocytes have been cultured or in liver perfusates (11). More recently, high-performance liquid chromatography (HPLC) has been used to quantify the relatively large increment in CE obtained with this substrate. Cholesteryl heptadecanoate is used as an internal standard and the assay avoids the use of a radioactive substrate.

The reagents required for the standard assay (10) are:

- Assay buffer—10 mM Tris–HCl (pH 7.4) containing 140 mM NaCl and 1 mM disodium EDTA
- Lecithin (ex. egg yolk)—10 mg/ml in redistilled $CHCl_3$-MeOH (9:1, v/v)
- UC—1 mg/ml in redistilled hexane
- [4-^{14}C]UC—55 µCi/µmol, 100 µCi/ml in toluene
- Human apo AI—1 mg/ml in assay buffer
- Sodium cholate—0.725 M in assay buffer

The cholate dialysis method for preparing the labelled proteoliposomes and the incubation conditions are shown in *Protocols 5* and *6* respectively.

Protocol 5. Preparation of labelled proteoliposome substrate by the cholate dialysis procedure

1. Pipette 870 µl of lecithin, 205 µl^a of UC and 20 µl^a of [^{14}C]UC into a clean tube and evaporate to dryness under N_2.

2. Add 1.7 ml of assay buffer, 1 ml of apo AI, and 340 µl of cholate solutions to the tube and vortex for 1 min at room temperature.

3. Dialyse against four daily changes (1 litre each) of assay buffer during 2–3 days at 4 °C to ensure complete removal of the inhibitory cholate.

Protocol 5. *Continued*

4. Adjust the volume of the dialysate to 4.5 ml with assay buffer; the substrate may be used immediately or stored in aliquots (0.1 ml) at −20 °C for up to 3 weeks.

[a] These volumes are suitable for measuring LCAT activity in plasma (about 10^5 c.p.m. per sample tube) but a 10-fold increase in the final [4-^{14}C]UC specific activity is required for assays of hepatocyte culture media.

Protocol 6. Incubation procedure with proteoliposomes as substrate

1. Pipette 0.235 ml assay buffer, 0.125 ml of 2% (w/v) HSA in assay buffer and 0.1 ml proteoliposome substrate (*Protocol 5*) into each screw-capped assay tube.
2. Pre-incubate at 37 °C for 20 min and then add 25 μl of 0.1 M mercaptoethanol and 15 μl of test plasma (or of assay buffer for blanks).
3. Mix and incubate at 37 °C for exactly 1 h. Stop the reaction by adding 3.75 ml of CHCl$_3$–MeOH (1:2, v/v), mix and then add 1.25 ml CHCl$_3$ and 1.25 ml H$_2$O.
4. Complete the TLC and radioactive counting steps as in *Protocol 3*.
5. Calculate the LCAT activity as in Section 2.1.1.*v* but in this case the UC content of the test plasma can be ignored (10) and the results may be expressed both as % esterification or nmol/ml/h.

2.3 Determination of LCAT concentration

A few laboratories have developed immunological assays for LCAT protein in plasma and plasma fractions, including a competitive double-antibody radioimmunoassay (7) and Laurell immunoelectrophoresis. Such assays avoid the problems associated with measuring UC esterification, although excellent correlation is reported between LCAT activity in normal plasma by the proteoliposome method (Section 2.2.1) and LCAT protein by radioimmuno-assay (7). Moreover, enzymatically-impaired or -inactive forms of LCAT may be present in abnormal plasma (e.g. from patients with familial LCAT deficiency) and these may still be detected by immunological methods. Unfortunately neither polyclonal nor monoclonal antibodies to LCAT are available commercially and, therefore, immunoassays are not yet in general use.

3. Acyl-CoA:cholesterol acyltransferase

ACAT is an integral protein of the rough endoplasmic reticulum and catalyses the intracellular formation of cholesteryl esters (*Figure 2*). The

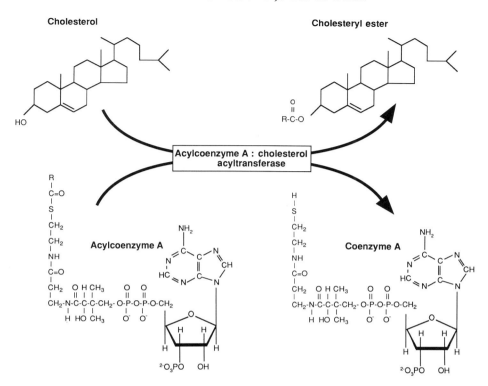

Figure 2. Reaction catalysed by acyl-CoA: cholesterol acyltransferase (ACAT)

metabolic role of this enzyme was initially studied in cultured skin fibroblasts, the model in which the LDL receptor pathway was first elucidated. Uptake of cholesterol by this route increases its rate of esterification by ACAT for storage and is accompanied by down-regulaton of receptor expression and decreased intracellular synthesis of UC. Many other cells have ACAT activity, including those with more complex fluxes of cholesterol than the fibroblast. In hepatocytes and enterocytes, CE synthesized by ACAT are incorporated into VLDL and chylomicra, respectively, and secreted. In adrenal cortex, and other steroidogenic tissues, the stored CE produced by ACAT can be hydrolysed by a hormone-sensitive CE hydrolase and the mobilized UC transferred to mitochondria where steroid hormone synthesis is initiated. Increasing evidence also implicates ACAT in depositing CE in arterial lesions, resulting in 'foam cell' formation. Measurement of ACAT activity in cells or subcellular fractions may be relevant, therefore, to studies on steroid hormone production, on lipoprotein synthesis and on new anti-atherogenic drugs.

3.1 Incorporation of labelled fatty acids into cellular CE

Goldstein *et al*. (12) first described incorporation of [1-^{14}C]oleic acid into CE following uptake of LDL by skin fibroblasts. Similar measurements have been made in other cultured cells, including parasitic organisms such as African trypanosomes which use receptor-mediated pathways to acquire lipids from the host. Note that the true substrate of ACAT, a fatty acyl ester of coenzyme A (CoA) is not used; unlike oleic acid it does not cross cellular plasma membranes. Thus, the assay is not an absolute measure of ACAT activity but rather the resultant activities of a transport process and two enzymatic reactions—conversion of oleic acid to its CoA ester and subsequent esterification to UC. Nevertheless, in fibroblasts at least, there is good agreement between incorporation of [^{14}C]oleic acid into CE and measurement of ACAT activity in microsomes from the same cells using [^{14}C]oleoyl-CoA as substrate.

The materials required for the assay are:

(a) Tissue culture reagents (from e.g. Gibco Ltd or ICN Flow Ltd)

  • Dulbecco's modification of Eagle's medium (DMEM) supplemented with penicillin (100 units/ml), streptomycin (100 µg/ml), and 10% (v/v) fetal calf serum (FCS).

  • Dulbecco's phosphate-buffered saline (PBS)

(b) Lipoproteins: Human LDL (1.019–1.063 g/ml) and lipoprotein-deficient serum (LPDS) are prepared as described in Chapter 1 and are extensively dialysed against 0.15 M NaCl containing 0.3 mM disodium EDTA and 10 mM phosphate buffer, pH 7.4; they are sterilized by passage through a 0.22 µm filter before use.

(c) [^{14}C]oleic acid/HSA emulsion:

  i. Pipette 50 µCi of [1-^{14}C]oleic acid (57 µCi/µmol) into a clean tube and remove the solvent under N$_2$. Alternatively 125 µCi of [9, 10-^{3}H]oleic acid, diluted to a specific activity of 25 mCi/mmol with unlabelled oleic acid, may be used.

  ii. Redissolve in 0.1 ml of 0.02 M NaOH by warming to 60 °C and vortexing to obtain a clear solution.

  iii. Transfer the hot solution dropwise to 0.9 ml PBS containing 1.5 mg HSA while vortexing. Transfer back to the original tube and vortex. The emulsion is then passed through a 0.22 µm filter and may be used immediately or stored at −20 °C for up to 1 month.

Protocol 7. Cell culture and incubation to incorporate labelled fatty acids into cellular CE

1. Grow cell in, for example, 60 × 15 mm plastic dishes containing supplemented DMEM (see under 'Tissue culture reagents' above) until they are in late log phase growth (i.e. nearly confluent).

2. Wash the cell monolayer twice with PBS and pre-incubate for 24 h with medium containing LPDS in place of FCS to maximize LDL receptor expression.

3. Replenish with fresh medium containing LPDS and varying amounts of test additive (e.g. LDL: 0–600 µg of LDL-cholesterol/ml) for 6 h and then add 40 µl [^{14}C]oleic acid/HSA emulsion.

4. After 2 h incubation, wash the cells twice with ice-cold PBS and harvest by scraping.

5. Transfer the cells in PBS to a glass tube, centrifuge (800g, 3 min, 4 °C) and lyse the pellet with 0.4 ml H_2O.

6. Whilst vortexing, add 3 ml of $CHCl_3$–MeOH (1:2, v/v) containing 10^5 d.p.m. of [7-^{3}H]UC and 40 µg each of unlabelled UC and cholesteryl oleate.

7. Centrifuge and, after transferring the supernatant to a clean tube, evaporate remaining traces of solvent before dissolving the protein residue (~400 µg) in exactly 1 ml 0.1 M NaOH for subsequent estimation.

8. Add 1 ml $CHCl_3$ and 1 ml H_2O to the supernatant, vortex and separate the [^{3}H]UC and [^{14}C]CE in the lower phase by TLC as in *Protocol 3*.

9. After counting, correct for recovery of the internal [^{3}H]UC (usually ~80%) for each dish and, using the specific activity of the [^{14}C]oleic acid, express the results as nmol CE formed per mg cell protein per 2 h.

3.2 Determination of ACAT activity in microsomes

Many of the methodological problems associated with direct assay of ACAT activity have been reviewed (13). Subcellular fractionation to prepare the microsomal fraction containing the enzyme is required. The preferred substrate is [^{14}C]oleoyl-CoA; palmitoyl-CoA is particularly sensitive to degradation by a second microsomal enzyme (acyl-CoA hydrolase), whilst use of [^{14}C]UC is complicated by its ill-defined equilibration with endogenous microsomal UC. However, the assay mixture must be supplemented with exogenous UC (added as cholesterol-rich liposomes or as a dispersion in detergent) for optimal ACAT activity and with a thiol to maintain the active –SH group of ACAT. Fatty-acid-free BSA is also included to prevent the detergent effect of oleoyl-CoA disrupting the microsomal membranes. The method detailed in *Protocols 8* and *9* is based on that of Bilheimer (14); although there are several published variations, this assay has proved reliable and applicable to such systems as rat ovary, insect tissues, and yeast.

The reagents required to assay ACAT activity are as follows:

(a) Assay buffer—0.1 M potassium phosphate (pH 7.4) containing 1 mM reduced glutathione

(b) BSA—20 mg/ml in assay buffer

(c) UC/Triton WR-1339 stock solution—33.4 mg UC and 1 g Triton WR-1339 in 10 ml acetone

(d) UC/Triton WR-1339 working solution. Make daily as follows for 10 samples: add 1 ml assay buffer to a screw-capped tube, weigh and warm to 60 °C. Warm stock solution to 60 °C to ensure that the UC is dissolved and then add 60 μl dropwise to the assay buffer whilst vortexing. Remove acetone under N_2 as described in *Protocol 1* and reweigh the tube to calculate the amount of lost water; replace this dropwise whilst vortexing. The resulting solution should be transparent (turbid solutions give falsely low results and must be discarded) and each ml contains 520 nmol UC and 6 mg Triton WR-1339 (final concentration in the assay mixture is 0.3%)

(e) [1-^{12}4C]oleoyl-CoA (5000 d.p.m./nmol)—1 mM in 0.01 M potassium phosphate (pH 6.0). Dilute the commercially available labelled oleoyl-CoA (50–60 μCi/μmol and stored under N_2 at −80 °C to avoid decomposition) with unlabelled oleoyl-CoA just before use.

Protocol 8. Preparation of microsomes

1. Perfuse extensively the tissue or organ *in situ* with ice-cold 0.25 M sucrose and homogenize in 2.5 vol of assay buffer containing 10 mM nicotinamide.

2. Use differential centrifugation (final centrifugation 105 000g, 1 h, 4 °C) to prepare the microsomes.

3. For use the same day, wash once with assay buffer. For storage under N_2 at −20 °C or lower, the microsomes should be washed with nicotinamide-containing buffer; after thawing wash again with assay buffer alone.

Protocol 9. Incubation procedure for ACAT activity

1. To each screw-capped tube, add 100 μg microsomal protein (boiled if a control), 50 μl BSA solution, and 100 μl UC/Triton WR-1339 working solution.

2. Complete to 0.18 ml with assay buffer and pre-incubate the tubes at 37 °C for 30 min.

3. Start the clock and, at 15 sec intervals, add 20 μl of [^{14}C]oleoyl-CoA to each tube and mix. After exactly 10 min stop the reaction in the first tube by adding 4 ml CHCl$_3$–MeOH (2:1, v/v) containing 40 μg of cholesteryl oleate and 40 000 d.p.m. of [^{14}C]UC as carrier and internal standard, respectively.

4. Repeat at 15 sec intervals for the other tubes.
5. Add 0.8 ml H$_2$O and separate the [^{14}C]CE and [^{14}C]UC in the lower phase by TLC as in section 2.1.1.*iv*.
6. Count and correct the radioactivity in CE for incomplete recovery of [^{14}C]UC. The activity of microsomal ACAT is expressed as nmol cholesteryl oleate formed per mg protein per min (typical values for rat liver ACAT are 0.4–0.6 nmol/mg protein/min).

References

1. Stokke, K. T. and Norum, K. R. (1971). *Scand. J. Clin. Lab. Invest.*, **27**, 21.
2. Yen, F. T. and Nishida, T. (1990). *J. Lipid Res.*, **31**, 349.
3. Freeman, D. J., Packard, C. J., Shepherd, J., and Gaffney, D. (1990). *Clin. Sci.*, **79**, 575.
4. Kaluzny, M. A., Duncan, L. A., Merritt, M. V., and Epps, D. E. (1985). *J. Lipid Res.*, **26**, 135.
5. Morris, L. J. (1963). *J. Lipid Res.*, **4**, 357.
6. Gillett, M. P. T. (1978). *Scand. J. Clin. Lab. Invest.*, **38** (Suppl. 150), 32.
7. Albers, J. J., Chen, C.-H., and Lacko, A. G. (1986). In *Methods in Enzymology* (ed. J. J. Albers and J. P. Segrest), Vol. 129, pp. 763–783. Academic Press, London.
8. Owen, J. S., Ramalho, V., Costa, J. C. M., and Gillett, M. P. T. (1979).*Comp. Biochem. Physiol.*, **63B**, 261.
9. Glomset, J. A. and Wright, J. L. (1964). *Biochim. Biophys. Acta*, **89**, 266.
10. Chen, C.-H. and Albers, J. J. (1982). *J. Lipid Res.*, **23**, 680.
11. Erickson, S. K. and Fielding, P. E. (1986). *J. Lipid Res.*, **27**, 875.
12. Goldstein, J. L., Dana, S. E., and Brown, M. S. (1974). *Proc. Natl. Acad. Sci. USA*, **71**, 4288.
13. Suckling, K. E. and Stange, E. F. (1985). *J. Lipid Res.*, **26**, 647.
14. Billheimer, J. T. (1985). In *Methods in Enzymology* (ed. J. H. Law and H. C. Rilling), Vol. 111, pp. 286–293. Academic Press, London.

$$\boxed{7c}$$

Assay of 3-hydroxy-3-methylglutaryl coenzyme A (HMG-CoA) reductase

PETER A. WILCE and PAULUS A. KROON

1. Introduction

The enzyme 3-hydroxy-3-methylglutaryl coenzyme A (HMG-CoA) reductase (EC 1.1.1.34 Mevalonate: $NADP^+$ Oxidoreductase (CoA Acetylating)) which catalyses the conversion of HMG-CoA to mevalonate is rate limiting in cholesterol biosynthesis. Although cholesterol is the major product formed, mevalonate is also an obligatory intermediate for the synthesis of a number of other isoprenoids such as: ubiquinone, a component of the electron transport system; dolichol, required for glycoprotein synthesis; and isopentyladenosine present in transfer RNA. Recently, an important growth regulatory function of mevalonate derived products has been demonstrated (reviewed in 1). The protooncogene p21^{ras} and proteins of the nuclear envelope have been shown to be attached to membranes by a convalently attached farnesyl residue. Inhibition of mevalonate synthesis prevents farnesylation of these proteins and blocks cell growth.

1.1 The enzyme

The conversion of HMG-CoA to mevalonate is essentially irreversible and occurs in two sequential reductive steps as shown in *Figure 1*. Enzyme-bound

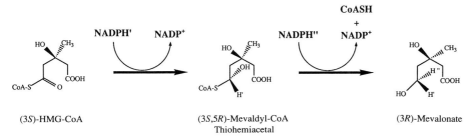

(3S)-HMG-CoA   (3S,5R)-Mevaldyl-CoA   (3R)-Mevalonate
Thiohemiacetal

Figure 1. The conversion of HMG-CoA to mevalonate.

thiohemiacetal of mevaldic acid and CoA are believed to be the intermediate products of the first reduction. Replacement of NADP$^+$ from the first step with NADPH results in the cleavage and reduction of the thiohemiacetal to produce mevalonate and CoA.

Most mammalian cells have the capacity to synthesize cholesterol and therefore contain the enzyme. HMG-CoA reductase has been studied in man and the traditional range of laboratory animals—rat, mouse, rabbit, guinea pig as well as in monkeys, gerbils, hamsters, pigs, dogs, and cattle. The list of tissues studied is as extensive. In man, activity has been studied in liver, intestine, corpus luteum, fetal adrenal gland, hair root, skin fibroblasts, and blood cells, while in rodents an even more diverse range of tissues have been studied including lung, brain, ovary, prostate gland, epididymal fat pad, and uterine epithelial cells.

1.2 Enzyme structure

Our understanding of the structure of HMG-CoA reductase protein has progressed quickly over the last decade. Early work in the seventies showed the enzyme to be a microsomal protein that could be solubilized from membranes by treatment such as freeze/thawing and incubation with high salt buffers (2). Purification of the soluble enzyme yielded a homogenous protein of about 53 000 Da (3). Subsequent studies have shown that this protein was a proteolytic fragment of the intact enzyme and that it constituted the catalytic domain. Analysis of the amino acid sequence predicted from its cDNA has suggested that the reductase has a unique structure (4). At the amino-terminus there are seven stretches of hydrophobic amino acids spanning the lipid bilayer. These regions are connected by short sections of less than 20 amino acids. The region between the sixth and seventh transmembrane regions is longer and reaches into the lumen of the endoplasmic reticulum. The carboxy-terminus contains the catalytic domain.

1.3 Regulation of the enzyme

In the presence of the cholesterol derived from LDL, HMG-CoA reductase activity is markedly reduced but is sufficient to provide mevalonate for non-sterol products. Maximal enzyme activity is achieved in cholesterol-depleted cells. The activity of HMG-CoA reductase can be regulated greater than 100-fold in response to cholesterol. Several mechanisms account for these large changes in enzyme activity. Cholesterol or its oxidized derivatives decrease the rate of transcription of the HMG-CoA reductase gene and decreases the half-life of the enzyme from 6 to 2 h. The exact mechanism that accounts for this increased rate of degradation is not clearly understood although the membrane domain is necessary. The degradation does not appear to involve ubiquitination or lysosomal mechanisms. The enzyme also appears to be under translational control (5). Other mechanisms have also been implicated in the regulation of the enzyme. These include inactivation by phosphorylation (6) and allosteric regulation by intracellular reducing agents (7).

2. Sample preparation for HMG-CoA reductase assays

Mammalian HMG-CoA reductase is located predominantly in the endoplasmic reticulum with the catalytic domain facing the cytoplasm. Activity measurements for tissues are generally performed on microsomes. A catalytically active and water-soluble protein fragment, prepared from microsomes, can also be used. HMG-CoA reductase measurements in cultured cells are most readily performed using detergent-solubilized whole cell extracts, although microsomes can also be used.

2.1 Important parameters

(a) *In vivo* HMG-CA reductase activity varies in a diurnal manner. In rats maintained on alternating 12-hour light–dark cycles, maximal activity occurs in the middle of the dark period. Reductase activity can vary up to 10-fold during this cycle. Appropriate timing is therefore critical.

(b) Reactions which compete for HMG-CoA should be minimized during the HMG-CoA reductase assay. Mitochondrial enzymes such as HMG-CoA lyase are particularly abundant in ketogenic organs such as liver and intestine, but can be avoided in homogenates from these organs by careful cellular fractionation.

(c) The solubilization of HMG-CoA reductase from microsomes appears to be mediated by lysosomal thiol proteinases (8). Both the soluble and microsomal enzymes are active, although there may be some differences in kinetic constants. If the intact microsomal enzyme is required it is important to isolate lysosome-free microsomes which have not been exposed to lysosomal proteolytic enzymes during the isolation procedure. To this end, leupeptin is included in the isolation buffer to inhibit thiol proteinases, and 250 mM sucrose is included to maintain the integrity of lysosomes and other organelles.

(d) HMG-CoA reductase is reversibly inactivated when cysteine residues are oxidized, and it has been argued that the activity of the enzyme within the cell depends on the intracellular thiol/disulphide status (7). To prepare microsomes which reflect the thiol/disulphide status of the cells from which they are isolated, reducing agents are not included in the isolation buffer.

(e) HMG-CoA reductase is reversibly inactivated by phosphorylation. Endogenous phosphatase can activate the enzyme during sample preparation (9). This can be avoided by adding phosphatase inhibitors such as fluoride at the time of homogenization.

2.2 Microsomal HMG-CoA reductase

In this section we describe the preparation of microsomes from rat liver. Similar procedures can be used for other tissues, and for cultured cells, although some variations will exist. For example, in brain homogenate, a large fraction of HMG-CoA reductase activity is recovered in low speed pellets, paralleling other microsomal markers (10). This probably occurs as a result of trapping or interactions with myelin.

Rats are sacrificed in accordance with institutional ethical procedures. Livers are immediately removed and placed in ice-cold homogenization buffer. *Protocol 1* outlines the procedures used to prepare microsomes. Isolation of microsomal HMG-CoA reductase by the method outlined in *Protocol 1* minimizes its susceptibility to proteolytic cleavage following freeze-thaw cycles. To isolate solubilized enzyme, microsomes are prepared in the absence of leupeptin. The solubilization procedure in *Protocol 2* is based on the method described by Heller and Gould (2) as modified by Edwards *et al.* (11).

Protocol 1. Preparation of microsomes from rat liver

1. On ice, mince liver with scissors into fine pieces.[a]
2. Wash briefly with homogenization buffer A.[b]
3. Suspend tissue in homogenization buffer (5 ml/g of liver).
4. Homogenize with a loose-fitting hand-operated Dounce homogenizer.
5. Centrifuge for 15 min at 16 000g.
6. Centrifuge the supernatant for an additional 15 min at 16 000g.
7. Carefully remove the upper three-fourths of the supernatant, avoiding the pellet.
8. Centrifuge the supernatant for 1 h at 100 000g.
9. Decant the supernatant and suspend the pellet in homogenization buffer A (2.5 ml/g of starting liver).
10. Centrifuge the suspended pellet for 1 h at 100 000g.
11. Decant the supernatant and drain residual liquid from the tube.
12. Microsomal pellets are frozen in liquid nitrogen and stored at −70 °C.

[a] All steps are performed at 0–4 °C.
[b] Homogenization buffer A contains 50 mM imidazole–HCl, pH 7.2, 250 mM sucrose, 20 mM EGTA, and 50 µM leupeptin.

Protocol 2. Solubilization of microsomal HMG-CoA reductase

1. Prepare microsomes as described in *Protocol 1* with the following changes. Use homogenization buffer B listed below,[a] and use a motor driven, tight-fitting, glass–Teflon Potter-Elvehjem homogenizer.

2. Suspend microsomes in homogenization buffer B at a concentration of approximately 80 mg/ml and add solid dithiothreitol to a final concentration of 10 mM.

3. Homogenize microsomes using a hand-operated, glass Potter-Elvehjem homogenizer (0.004 to 0.006″ clearance). In subsequent steps the same homogenizer is used.

4. Freeze 3 ml aliquots to −20 °C in glass tubes at a rate of 6–8 °C per min. Samples can be stored at −20 °C.

5. At the time of analysis microsomal HMG-CoA reductase is solubilized. Thaw samples at 37 °C and add an equal volume of preheated (37 °C) homogenization buffer B containing 50% glycerol and 10 mM dithiothreitol.

6. Homogenize with 10 downward passes.

7. Incubate at 37 °C for 1 h.

8. Dilute with two volumes of preheated (37 °C) homogenization buffer B containing 10 mM dithiothreitol.

9. Homogenize with 10 downward passes.

10. Centrifuge for 1 h at 100 000g, and 25 °C.

11. Remove the supernatant which contains solubilized reductase. This can be used directly in the spectophotometric HMG-CoA reductase assay, or for further purification.

[a] Homogenization buffer B consists of 100 mM sucrose, 50 mM KCl, 40 mM potassium phosphate, and 30 mM potassium EDTA, pH 7.2.

2.3 HMG-CoA reductase from cultured cells

HMG-CoA reductase activity in cultured cells is measured on detergent solubilized whole cell extracts, although microsomes can also be used. Two steps are involved: (i) cells are harvested and frozen following an experiment, and (ii) just before analysis, a detergent extract is prepared. It should be noted that significant activity is only present in cultured cells after incubation in lipoprotein-deficient serum for approximately 18 h.

For cells grown in suspension: centrifuge at 900g for 3 min, room temperature, and suspend in 50 mM Tris–HCl, pH 7.4, 150 mM NaCl (1 ml/ 2×10^6 cells). Centrifuge again and wash the pellet once more in the same manner. Pellets are frozen in liquid nitrogen and stored at −70 °C until use.

For attached cells: discard the medium from Petri dishes (35 mm or 60 mm), and scrape cells with a rubber policeman into 1 ml of 50 mM Tris–HCl, pH 7.4, 150 mM NaCl. Centrifuge cells at 900g, for 3 min at room temperature, and wash the pellet once more in the same manner. Freeze and store pellets as for suspension cells.

3. Measurement of HMG-CoA reductase activity

The most commonly used method to measure HMG-CoA reductase activity is based on the formation of [3-^{14}C]mevalonate from [3-^{14}C] HMG-CoA. This method is sensitive, and is the method of choice for membrane-bound HMG-CoA reductase. Following the reaction, the precursor and product must be separated to allow quantitation of the product formed. This is the most time consuming part of the assay. Thin-layer, column, and paper chromatographic procedures are described to achieve this separation. [^{3}H]mevalonate is included prior to chromatography to correct for procedural losses. This assay can be used for both the microsomal enzyme and for the soluble 53 kDa catalytic domain.

A spectrophotometric method is also described. This method measures the HMG-CoA reductase-dependent oxidation of NADPH to NADP$^+$ (11). This method is only suitable for the soluble catalytic domain of reductase preparations; background reactions and light scattering make this method unsuitable for microsomes. Advantages of this assay are its speed (reactants do not have to be separated), and the fact that no radioactive substrates are used. This method can be used effectively to investigate compounds which modulate the activity of HMG-CoA reductase.

3.1 Important parameters

(a) Apparent K_m values for (S)-HMG-CoA are in the range of 1 to 21 μM for the rat liver microsomal enzyme, and appear to depend on the dietary treatment of the animal (7, 12). The microsomal enzyme exhibits sigmoidal kinetics when the NADPH concentration is varied. NADPH concentrations required to achieve 50% of the maximal velocity ($S_{0.5}$) depend on the dietary treatment, varying from 40 μM for rats fed a diet containing lovastatin and cholestipol to 1.3 mM for those fed a diet contained cholesterol (13). Preincubation of HMG-CoA reductase with NADPH prior to analysis of the enzyme activity is standard practice; this results in a 4 to 6-fold increase in $S_{0.5}$ values.

(b) Competing reactions are minimized by using phosphate buffer, which inhibits cleavage enzymes (14). In some tissues such as liver, the presence of high activity in homogenates, low speed pellets, and low speed supernatants makes the assay of reductase in these fractions unreliable.

To determine whether cleavage activity or other factors diminish the production of mevalonate, microsomes prepared from rat liver and the whole cell homogenate of interest are assayed separately and together. If the amount of mevalonate produced by the mixture is less than that products by the individual reactions, a contaminating activity (or inhibitor) is present.

(c) HMG-CoA reductase activity is modulated by reversible oxidation–reduction of sulphydryl groups. In a glutathione redox buffer, rat liver HMG-CoA reductase rapidly equilibrates between a reduced active form, and an oxidized inactive form (15). For maximal activity a thiol such as dithiothreitol is therefore required in the assay. If thiols are omitted the activity will more closely reflect the thiol/disulphide status of the cell from which the enzyme was isolated.

(d) HMG-CoA reductase activity is modulated by reversible phosphorylation. Phosphorylation severely impairs catalytic activity which is completely restored upon treatment with fluoride-sensitive phosphatases. Measurement of 'expressed' activity of native HMG-CoA reductase, reflected by the degree of phosphorylation of the enzyme, and of 'total' activity of the dephosphorylated (protein phosphatase-treated) enzyme yields the percentage of active enzyme.

(e) For each tissue for which the HMG-CoA reductase activity is to be measured, linearity of the reaction with respect to concentration and time must be established.

(f) The protein concentration of soluble enzyme can be measured by any of the usual methods for protein estimation. If dithiothreitol is used in the preparation of microsomes, interference with the protein assay must be avoided. To prevent light scattering when membrane protein concentration is determined, an assay in which membranes are solubilized such as the Peterson modification of the Lowry assay (16) should be used.

(g) One unit of enzyme activity is defined as the amount required for the synthesis of 1 nmol of mevalonate per minute. For the spectophotomeric assay this corresponds to the conversion of 2 nmol of NADPH to $NADP^+$.

3.2 Radiochemical assay

The radiochemical assay uses (R, S)-$(3\text{-}^{14}C]$ HMG-CoA as a substrate. The (S)-isomer is reduced to mevalonate with the concommitant oxidation of 2 moles of NADPH to $NADP^+$. A constant concentration of NADPH is maintained during the reaction by utilizing the $NADP^+$-dependent oxidation of glucose 6-phosphate by glucose 6-phosphate dehydrogenase. The total reaction volume for the assay is 100 μl which contains 100 mM potassium phosphate, pH 7.4; 20 mM glucose 6-phosphate; 2.5 mM NADP; 1 unit of glucose 6-phosphate dehydrogenase; 5 mM dithiothreitol; and 1 mM EDTA. The procedure for this assay is outlined in *Protocol 3*.

Protocol 3. Radiochemical assay of HMG-CoA reductase

1. Sample preparation:

   (a) Microsomes—thaw frozen liver microsomal pellet and disperse in 200 mM phosphate buffer C[a] by passage through syringe needles of decreasing gauge.

   (b) Cell free extract—thaw fibroblast cell pellet and dissolve in 200 μl of 50 mM phosphate buffer D.[b]

2. Determine the protein concentration of the enzyme preparation.[c]

3. To obtain a final assay volume of 100 μl combine the following in a 1.5 ml microcentrifuge tube on ice:

   • 10 μl of a solution containing 25 mM NADP and 200 mM glucose 6-phosphate

   • Microsomes: 60–240 μg protein in phosphate buffer to a total volume of 50 μl; or cell free extract: 50–100 μg protein in 40 μl, and 200 mM phosphate buffer, 40 μl

   • 1 unit of glucose 6-phosphate dehydrogenase

   • water to bring the volume to 95 μl

4. Incubate at 37 °C for 15 min.

5. Add 5 μl of the substrate, (R,S)-[3-^{14}C]HMG-CoA (8 Ci/mol, 1.0 mM),[d] to give a final concentration of 50 μM.

6. Incubate at 37 °C for periods ranging from 5 to 120 min depending on the reductase activity.

7. The reaction is terminated either by (a) the addition of 10 μl of 5 M HCl if products are analysed by TLC (see Section 3.2.1) or descending paper chromatography (see Section 3.2.2) or by (b) the addition of 10 μl of 33% KOH if the products are to be analysed by column chromatography (see Section 3.2.3).

8. [5-^{3}H]mevalonolactone (0.01 μCi) is added together with unlabelled mevalonolactone (1 mg) as an internal standard and carrier for subsequent purification steps. The samples are incubated for an additional 30 min at 37 °C. During this time, samples treated with HCl lactonize.

9. Acidify samples treated with KOH in step 7 by adding 5 μl of 0.05% bromophenol blue, followed by 5 M HCl until the colour changes to yellow (20–25 μl), and incubate for 30 min to complete conversion of mevalonate to mevalonolactone.

Protocol 3. *Continued*

[a] Phosphate buffer C contains 200 mM K_2HPO_4, 2 mM EDTA, and 10 mM dithiothreitol at pH 7.4.

[b] Phosphate buffer D contains 50 mM K_2HPO_4, pH 7.4, 5 mM dithiothreitol, 1 mM EDTA. Some workers add a synthetic nonionic detergent such as Kyro EOB to this buffer at a concentration of 0.25%. Kyro EOB is no longer available from the original suppliers but Tergitol-15-5-9 has a similar structure (D. H. Hughes, Proctor and Gamble Miami Valley Research Laboratories, personal communication) and can be obtained from Union Carbide. However, there are no reports of its use in the literature to date.

[c] Protein assay is performed using the Peterson modification of the Lowry assay (16).

[d] Substrate solution contains 200 μl [3-^{14}C]HMG-CoA (20 μCi); 46 μl of HMG-CoA (10 mg/ml); and 254 μl H_2O.

Effective separation of labelled precursor, HMG-CoA from the product of the reaction, mevalonolactone, is critical. We describe three chromatographic procedures to achieve this.

3.2.1 Thin-layer chromatography

Separation of HMG-CoA and mevalonolactone by thin-layer chromatography is effective, and is still the most commonly used method. Before analysis, denatured protein is pelleted by centrifugation (5 min) in a microcentrifuge. Aliquots (75 μl) are applied to 2.0 cm wide lanes on activated silica gel G thin-layer chromatographic plates M (750 μm thickness, Merck), and developed in acetone/benzene (1:1 v/v). Alternatively, the plate can be developed with chloroform/acetone (2:1, v/v). The plate is dried, and the mevalonolactone located (R_F = 0.7) using a radiochromatogram scanner or by staining with iodine vapour. The area containing mevalonolactone is scraped from the plate into a counting vial. Scintillation fluid (10 ml) is added, swirled to disperse the silica gel and dissolve the mevalonolactone, and the amount of radioactivity measured using a scintillation counter. Specific activity of the enzyme is calculated after correcting for counting efficiency and recovery of mevalonolactone from the silica gel.

3.2.2 Paper chromatography

This method is ideally suited for experiments where large numbers of samples are generated (17). The method is based on the observation that mevalonic acid lactone is soluble in toluene (18, 19) which allows the enzyme product to be easily separated by descending chromatography. A large chromatography tank 250 × 400 × 250 mm with two reservoirs is fitted with a rack to take four rows of 14 scintillation vials (20 ml). The tank is set up for descending chromatography using four combs of Whatman 3 mm chromatography paper cut into 14 fingers such that the tip of the finger fits well into a scintillation vial. The acidified reaction mix is poured onto the central region of each finger of the paper comb (*Figure 2*). The paper is dried thoroughly (60 °C, 15 min) and the mevalonolactone eluted with toluene-based scintillation fluid

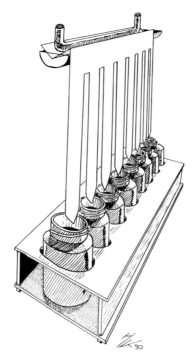

Figure 2. Diagrammatic representation of the descending chromatography apparatus. Acidified reaction mixture is loaded on the central part of the paper fingers and the paper thoroughly dried. Mevalonolactone is eluted with scintillation fluid. Only eight fingers are shown in the diagram, however 14 are conveniently cut from a standard sheet of chromatography paper.

(4 g/litre 2,5-diphenyloxazole, 50 mg/litre 1,4-bis [2-(4-methyl-5-phenyl-oxazolyl)]benzene) directly into scintillation vials. An elution volume of 10 ml extracts 70% of the product and takes about 3 h. Increasing the elution volume to 20 ml increases recovery to 77%.

3.2.3 Column chromatography

This method provides a more rapid alternative to thin-layer chromatography (20). The reaction is stopped with KOH (see above) to hydrolyse unreacted [^{14}C]HMG-CoA. This results in a significant reduction in background counts when the column procedure is used (7). The lactonized sample is centrifuged for 5 min in a microcentrifuge to remove precipitated protein. AG1-X8 formate (200–400 mesh; Bio-Rad) columns (7 × 100 mm) are prepared in water. The supernatant is applied to the column and is eluted with water. The first 1.8 ml is discarded, and the mevalonolactone collected in the next 5 ml fraction. Four ml of this fraction is mixed with 15 ml of Aquasol (Beckman) and counted. Specific activity of the enzyme is calculated after correcting for counting efficiency and recovery of mevalonolactone from the column.

3.2.4 Other methods of separation

Several other methods for separation of product from substrate have been described. Goodwin and Margolis (21) described extraction with benzene. The mevalonolactone is formed by incubation of the assay mixture in a final concentration of 6 M HCl for up to 60 min. The pH is then returned to approximately 6.5 by the addition of saturating amounts of sodium sulphite (0.3 g per 100 µl assay). Mevalonolactone is isolated by two successive extractions with 7.5 ml of benzene. Aliquots of the benzene are then taken, evaporated in scintillation vials, and counted in an organic scintillant. The original authors report good agreement with the thin-layer method described above, both for extraction efficiency and background radioactivity. Benzene, however, is carcinogenic and it is not stated whether other solvents such as toluene can be substituted.

A simple mixed phase assay in which mevalonolactone is extracted directly into a toluene-based scintillation fluid has been described by Philipp and Shapiro (18). Ackerman *et al.* (22) described a chloroform extraction procedure. The assay mixture (made up to 1 ml) is first incubated at 37 °C for 30 min in the presence of 0.5 g of solid equimolar Na_2HPO_4/NaH_2PO_4. This mixture is subsequently extracted overnight with chloroform. The chloroform is separated by Whatman IPS phase-separating filter paper and evaporated. The mevalonolactone is dissolved in scintillation fluid and counted.

The three methods described above are fairly time consuming and use large quantities of hazardous solvents. They do however have good efficiencies of product extraction.

3.3 Spectophotometric assay of HMG-CoA reductase

The activity of the solubilized enzyme can be determined by monitoring the rate at which the absorbance at 350 nm decreases due to the oxidation of NADPH. The conditions below are for a 1 ml cuvette with a 1 cm lightpath. Microcuvettes with a 1 cm lightpath can be used to scale the reaction down to smaller volumes. The reaction mixture contains:

- 600 µl of phosphate buffer (300 mM KCl, 240 mM potassium phosphate, 6 mM EDTA, and 15 mM dithiothreitol, pH 6.8)
- 100 µl 2 mM NADPH
- 100 µl 1 mM (*R,S*)-HMG-CoA
- 100 µl of solubilized reductase
- 100 µl of H₂O or inhibitor, stimulator etc.

The rate of oxidation of NADPH is initially determined in the absence of HMG-CoA and this blank value subtracted from the rate obtained with both substrates.

3.4 Other methods of assay

Other methods of assay have been described. One such method measures the CoASH released in the action either by a colorimetric reaction (23) or by enzymatic determination with 2-oxoglutarate dehydrogenase (24). The former method has been reported to generate a high background. The latter method seems only reliable with enzyme activities above 0.2 nmol mevalonolactone/min/mg protein but has the advantage of being a fluorometric method and therefore fairly rapid and cost effective. The enzyme 2-oxoglutarate dehydrogenase is not commercially available but can be readily purified from pig heart.

References

1. Goldstein, J. L. and Brown, M. S. (1990). *Nature*, **343**, 425.
2. Heller, R. A. and Gould, R. G. (1973). *Biochem. Biophys. Res. Commun.*, **50**, 859.
3. Edwards, P. A., Lemongella, D., Kane, J., Schechter, I., and Fogelman, A. M. (1980). *J. Biol. Chem.*, **255**, 3715.
4. Liscum, L., Finer-Moore, J., Stroud, R. M., Luskey, K. L., Brown, M. S., and Goldstein, J. L. (1985). *J. Biol. Chem.*, **260**, 522.
5. Nakanishi, M., Goldstein, J. L., and Brown, M. S. (1988). *J. Biol. Chem.*, **263**, 8929.
6. Beg, Z. H. and Brewer, H. B., Jr (1981). *Curr. Top. Cell. Regul.*, **20**, 139.
7. Roitelman, J. and Schecter, I. (1984). *J. Biol. Chem.*, **259**, 870.
8. Chin, D. J., Luskey, K. L., Anderson, R. G. W., Faust, J. R., Goldstein, J. L., and Brown, M. S. *Proc. Acad. Sci. USA*, **79**, 1185.
9. Gibson, D. M. and Ingebritsen, T. S. (1978). *Life Sci.*, **23**, 2649.
10. Sudjic, M. M. and Booth, R. (1976). *Biochem. J.*, **154**, 559.
11. Edwards, P. A. Lemongella, D., and Fogelman, A. M. (1979). *J. Lipid Res.*, **20**, 40.
12. Ness, G. C., Sample, C. E., Smith, M., Pendleton, L. C., and Eichler, D. C. (1986). *Biochem. J.*, **233**, 167.
13. Ness, G. C., McCreery, M. J., Sample, C. E., Smith, M., and Pendleton, L. C. (1985). *J. Biol. Chem.*, **260**, 16395.
14. Stegink, L. D. and Coon, M. J. (1968). *J. Biol. Chem.*, **243**, 5272.
15. Cappel, R. E. and Gilbert, H. F. (1988). *J. Biol. Chem.*, **263**, 12204.
16. Peterson, G. L. (1977). *Anal. Biochem.*, **83**, 346.
17. Gregg, R. G. and Wilce, P. A. (1985). *Intl. J. Biochem.*, **17**, 707.
18. Philipp, B. W. and Shapiro, D. J. (1979). *J. Lipid Res.*, **20**, 588.
19. Murthy, H. R., Lupien, P. J., and Moorjani, S. (1978). *Anal. Biochem.*, **89**, 14.
20. Avigan, J., Bhalthena, S. J., and Schreiner, M. E. (1975). *J. Lipid Res.*, **16**, 151.
21. Goodwin, C. D. and Margolis, S. (1976). *J. Lipid Res.*, **17**, 297.
22. Ackerman, M. E., Redd, W. L., Tormanen, C. T., Hargrave, J. E., and Scallen, T. J. (1977). *J. Lipid Res.*, **18**, 404.
23. Hulcher, F. H. and Olsen, W. H. (1973). *J. Lipid Res.*, **14**, 625.
24. Baqir, Y. A. and Booth, R. (1977). *Biochem. J.* **164**, 501.

Analysis of tissue lipoproteins

ROBERT E. PITAS and ROBERT W. MAHLEY

1. Introduction

This chapter describes in detail procedures we have used to analyse the lipoproteins present in cerebrospinal fluid (1) and tissue extracts (2) and to detect apolipoproteins in tissues (2). These procedures include extraction of apolipoproteins and lipoproteins from tissue and their subsequent analysis, as well as the detection of apolipoproteins within tissues and cells by immunocytochemistry. The coverage is not meant to be all-inclusive but rather to describe techniques that are in routine use in our laboratory and that may be applied to other systems.

2. Extraction of apolipoproteins and lipoproteins from tissue

Contamination by plasma proteins or lipoproteins can be avoided by whole-body perfusion of the animal with ice-cold tissue culture medium or phosphate-buffered saline (PBS) (2), by inserting a catheter into the left ventricle of the heart of the anesthetized animal, cutting the inferior vena cava, and perfusing the animal with medium (250 ml minimum for a rat) until the outflow is free of blood. The liver changes to a lighter colour as the blood is cleared. The method of extraction after perfusion will depend on whether intact lipoproteins or apolipoproteins are to be studied.

2.1 Extraction of intact lipoproteins

Care must be taken to ensure that the lipoproteins are not denatured or altered in composition. The lipoproteins are leached from the tissue into ice-cold extraction buffer (2.05% Tris–HCl, 0.88% NaCl, 0.05% EDTA, pH 7.4) containing protease inhibitors but no detergents. This procedure has been used to extract intact lipoproteins from normal and regenerating nerves (3) and from atherosclerotic lesions (4). Cut the tissue into small pieces and then incubate them with the buffer (10 to 20 ml/g of tissue) containing protease inhibitors and preservatives (0.13 g of ε-amino caproic acid, 2 mg of chloramphenicol, 0.2 mg of leupeptin, 0.01 ml of aproteinin, 2.4 mg of pepstatin A, 50 000 units of penicillin G, and 5 mg of streptomycin sulphate per 100 ml of buffer) for 16 h at 4 °C. The amount of time necessary to obtain maximum extraction is determined empirically. Concentrate the lipoproteins by adjusting the density of the solution to 1.21 g/ml with solid KBr and then isolate the lipoproteins by centrifuging the solution (see Section 3).

2.2 Extraction of apolipoproteins

Apolipoproteins are extracted by homogenization of the tissue in the presence of detergent to obtain intercellular and intracellular apolipoproteins. Tissues are homogenized (Polytron homogenizer, Brinkman Instruments) by three 15-sec pulses at 4 °C at a setting of 8. If the extract is to be applied directly to gels for separation and Western blot analysis (5), tissues from perfused animals can be homogenized directly in SDS–PAGE sample application buffer (see *Protocol 9*), and the low-speed supernatant analysed. It is often useful, however, to immunoprecipitate the protein of interest from the homogenate prior to analysis (see Section 4.2). In this case the tissue is homogenized in detergent-containing buffer (50 mM Tris–HCl, pH 8.5, containing 120 mM NaCl and 0.5% Nonidet-P40 (NP-40)) (10 ml/g of tissue), and the apolipoproteins in the $100\,000g$ supernatant are immunoprecipitated (2).

3. Isolation of lipoproteins by ultracentrifugation

Lipoproteins are separated from contaminating proteins by their buoyant density. Plasma lipoproteins are often isolated by sequential density ultra-centrifugation (6) because their densities are known. However, for the analysis of lipoproteins with unknown compositions, it is best to isolate the entire lipoprotein fraction for analysis first and then to separate the lipoproteins by density gradient ultracentrifugation to determine the true density ranges of the lipoproteins (7).

3.1 Isolation of total lipoprotein fraction

For isolation of the total lipoprotein fraction, the density of the solution containing the lipoprotein is raised to 1.21 g/ml. The density of body fluids such as plasma, cerebrospinal fluid, peripheral lymph, and lipoproteins extracted from tissues is approximately 1.006 g/ml. Raise the density to 1.21 g/ml by adding solid KBr (0.3265 g/ml), and isolate the lipoproteins by centrifugation (48 h at $360\,000g$ and 4 °C). Remove the lipoprotein fraction from the top of the tube by tube slicing. The lipoprotein fraction can be washed by diluting it with density solution ($d = 1.21$ g/ml) and recentrifuging at $360\,000g$, but for 24 h. Dialyse the lipoproteins against 0.15 M NaCl containing EDTA (saline-EDTA) (1 mg/ml), pH 7.0–7.4, to remove the KBr.

3.2 Subfractionation of lipoproteins by density gradient ultracentrifugation

Density gradient ultracentrifugation is performed by raising the density of the solution containing the lipoproteins to 1.21 g/ml as described above, overlaying this solution with a discontinuous density gradient, and centrifuging it in a swinging bucket rotor (7).

3.2.1 Preparation of density solutions

Prepare the 1.006 g/ml solution by dissolving 11.4 g of NaCl and 0.1 g of EDTA in 500 ml of water, adding 250 μl of 1 M NaOH, adjusting the pH to 7.0, and bringing the volume to 1 litre. Prepare other density solutions by adding solid KBr to this stock solution. The general formula for the preparation of the density solutions is: KBr (g) $= V_i (D_f - D_i)/(1 - \bar{v}D_f)$ where V_i is the volume of the solution to be adjusted (i.e. initial volume), D_f is the final density in g/ml, D_i is the initial density in g/ml, and $\bar{v}$ is the partial specific volume of KBr at the final density. The partial specific volume of KBr, which varies with concentration, has been reported (8). To prepare density solutions of 1.02, 1.05, 1.063, 1.1, 1.125, or 1.21 g/ml, respectively, add 0.0199, 0.0635, 0.0834, 0.1405, 0.1802, or 0.3265 g of solid KBr per ml of the 1.006 g/ml stock solution. The density of the solutions should be checked gravimetrically.

3.2.2 Preparation of centrifuge tubes

To layer density solutions, the centrifuge tubes must have a 'wettable' inner surface that allows the fluid to flow gently down the side of the tube. Beckman ultraclear and polyallomer tubes (Beckman Instruments), can be treated as follows to make them wettable (9):

(a) In a 250 ml round-bottom flask, dissolve 2 g of polyvinyl alcohol in 50 ml of water by stirring and heating to gentle reflux. Slowly add 50 ml of isopropanol while continuing to heat and mix until the solution is clear. Cool the solution to room temperature.

(b) Fill the centrifuge tubes with the polyvinyl alcohol solution and allow them to stand for 15 min. Aspirate the solution from the tubes. The material that collects in the bottom is removed by two subsequent aspirations. Air-dry the tubes overnight, fill with distilled water, leave overnight, and then wash five times with distilled water. Shake the tubes to remove excess water, and air-dry them.

Polyallomer tubes are better for density gradient centrifugation because they are easier to puncture for sample collection.

3.2.3 Density gradient centrifugation

(a) Adjust the lipoprotein solution to a density of 1.21 g/l with solid KBr, and place 4 ml of the solution in the bottom of an ultracentrifuge tube for an SW 41 rotor (Beckman). Overlay with 3.0 ml of 1.063 g/ml solution, 3.0 ml of 1.02 g/ml solution, and fill to the top with 1.006 g/ml solution (2.5 to 3.0 ml). Centrifuge at 41 000 r.p.m. for 18 h at 4 °C, and collect the lipoproteins by puncturing the bottom of the tube and collecting 0.5- to 0.75-ml fractions. We use a Beckman Fractionator. During the centrifugation, the gradient smoothes out and does not retain sharp density divisions (7).

(b) Determine the absorbance of each fraction at 280 nm, to locate protein, and the density of each fraction from the refractive index. The refractive index is related to the density of solutions containing KBr by the following formula. Density (g/ml) = (refractive index) (6.4786) − 7.6431. The lipoproteins and their apolipoproteins can be characterized as described below.

4. Analysis of the apolipoproteins and lipoproteins present in tissue extracts

4.1 Heparin–Sepharose column chromatography

Lipoproteins that contain apo E or apo B can be separated readily from other lipoproteins on the basis of their binding to heparin. This is conveniently done by passing lipoproteins through a column of heparin bound to Sepharose. The lipoproteins that do not contain apo B or apo E do not bind to the column, whereas apo B- and apo E-containing lipoproteins are retained and can be eluted at high salt concentrations. We have used this procedure to separate cerebrospinal fluid lipoproteins based on their apo E content (1). The preparation of the heparin–Sepharose has been described (10). The size of the column to be used depends entirely on the amount of apo B- and apo E-containing lipoproteins in the sample; 1 ml of heparin–Sepharose (hydrated volume) prepared as described will bind 3 mg of lipoprotein protein. For separation of large amounts of lipoproteins, refer to the original manuscript (10). For the separation of the apo E- and apo AI-containing lipoproteins in cerebrospinal fluid, *Protocol 1* was followed.

Protocol 1. Separation of lipoproteins in cerebrospinal fluid

1. Add heparin–Sepharose (1 ml hydrated volume) to a Pasteur pipette plugged with glass wool. Wash the column with 10 ml of cold sample application buffer (10 mM Tris, pH 7.4 containing 25 mM NaCl and 25 mM MnCl$_2$). Maintain the column and buffers at 4 °C. Dialyse the sample in the sample application buffer, and apply it to the column in a volume of 100 µl containing ~150 µg of lipoprotein protein. If necessary,

concentrate the sample prior to application using a Centricon 30 concentrator (Amicon Corp.).

2. Equilibrate the sample on the column for 30 min at 4 °C, and then elute with column application buffer to collect non-bound lipoproteins. The non-bound lipoproteins are usually eluted in the first 1.5 ml. Elute the bound fraction with 1.5 ml of elution buffer (10 mM Tris, pH 7.4, containing 500 mM NaCl).

3. Dialyse the lipoproteins extensively against saline-EDTA, and determine the protein concentration by a modification of the Lowry protein assay (11) (see *Protocol 2*). It is essential that the lipoproteins are dialysed free of the column buffers since the Mn^{2+} interferes with the determination of protein concentration.

4. If the samples contain both apo E- and apo B-containing lipoproteins, a different procedure is followed. Apply the samples to the column, and collect the non-bound fraction as described above. Elute the column with 10 mM Tris–HCl (pH 7.4) containing 95 mM $NaCl_2$ until the apo E-containing peak is collected. Then increase the salt concentration to 500 mM NaCl and elute the apo B-containing lipoproteins. The salt concentrations and the amount of each buffer required to elute the apo E- and apo B-containing lipoproteins separately must be determined empirically (10). The exact salt concentrations and volumes will vary to some extent for different batches of heparin–Sepharose and will also vary depending upon the species from which the lipoproteins were derived.

5. Dialyse or lyophilize the sample as appropriate prior to subsequent analysis.

4.1.1 Protein determination

The following procedure for determining small amounts of protein is a sensitive modification of that described by Lowry *et al.* (11). The assay is linear from < 0.5 to 20 µg of protein.

Protocol 2. Procedure for protein determination

1. Prepare reagents: Reagent A, 3.3% Na_2CO_3 in 0.06 M NaOH; Reagent B_1, 2% sodium tartrate; Reagent B_2, 1% $CuSO_4$. These reagents are stable at room temperature.

2. Prepare Reagent C immediately prior to use by mixing 50 ml of Reagent A with 0.8 ml of Reagent B_1, and then mix with 0.8 ml of Reagent B_2.

3. Aliquot samples and standards (bovine serum albumin, 0 to 20 µg). The volume must be 200 µl or less. Adjust the volume of each to 200 µl by adding 0.1 M NaOH.

Protocol 2. *Continued*

4. Add 0.33 ml of Reagent C, vortex, and let stand for 10 min. Add 27.5 μl of Folin and Ciocalteu's phenol reagent (2.0 N) while vortexing, and incubation for 40 min in the dark.

5. Record the absorbance at 750 nm.

4.2 Immunoprecipitation

Apolipoproteins can be immunoprecipitated from tissue extracts using specific polyclonal antiserum and *Staphylococcus aureus* Protein A (IgGSORB, The Enzyme Center). A 10% cell suspension is prepared as described in the package insert for the IgGSORB.

Protocol 3. Immunoprecipitation of apolipoproteins

1. Preclearing of non-specific precipitate:

(a) Add 50 μl of non-immune serum per 1 ml of supernatant obtained from tissue extract at 100 000g. The non-immune serum should be from the species in which the primary antibody was prepared.

(b) Incubate at 4 °C for 90 min with occasional gentle mixing.

(c) To precipate non-specific complexes, add 200 μl of a 10% suspension of *S. aureus*; mix gently on a rocking platform at room temperature for 10 min.

(d) Pellet the complexes by centrifugation (5 min at 2000g or 1 min at 10 000g). Discard the pellet.

2. Preparation of immunoprecipitant:

(a) Transfer aliquots of the precleared extract into two centrifuge tubes. Add 10–20 μl of pre-immune serum to one and the same volume of immune serum to the other.

(b) Repeat steps 1(b) and 1(c), and pellet the complexes by centrifugation. Discard the *supernatant* by aspiration.

(c) Wash the pellet twice as follows: suspend it in extraction buffer (Section 2.2) by sonication in a sonication bath (Bransonic 220, Branson Ultrasonics), for 10 to 15 min at 4 °C, vortex, centrifuge it, and discard the supernatant.

3. Preparation of samples for SDS–PAGE:

(a) Add 50 μl of SDS–PAGE sample application buffer (pH 6.8, see *Protocol 9*) to each pellet.

(b) Vortex, sonicate as described above to resuspend, heat at 100 °C for 3 min in an uncapped tube, and centrifuge to pellet the *S. aureus* Protein A.

(c) Samples (35–45 µl) are ready to be loaded into wells of SDS-PAGE gels. Samples that must be temporarily stored (−70 °C) should always be reheated to 100°C for 3 min prior to use to ensure that the complexed proteins are completely freed from the Protein A.

4.3 Immunoaffinity column chromatography

If specific antibodies are available, lipoproteins can be separated on the basis of their apolipoprotein content by affinity chromatography (1, 12). Affinity columns made by attaching anti-apolipoprotein immunoglobulin G (IgG) to Sepharose bind lipoproteins that contain these apolipoproteins.

Protocol 4. Isolation of immunoglobulin G from serum or ascites

Isolate IgG by applying the serum or ascites to a column of Protein A–Sepharose (Pharmacia), essentially as described (13). Perform all steps at 4 °C.

1. Wash column with 10–20 ml of elution buffer (0.1 M citric acid, adjust to pH 3.0 with 5 M NaOH).

2. Wash column with running buffer (1.5 M glycine, 3 M NaCl; adjust to pH 8.9 with 5 M NaOH).

3. Dilute ascites or serum 1:1 (v/v) with running buffer. Recycle the sample over the column overnight at ~10 ml/h.

4. Wash the column with 0.8 ml of running buffer/min until the absorbance at 280 nm is less than 0.10. Discard this wash.

5. Reverse the flow through the column. Elute the IgG from the column with elution buffer, collecting 1.5 ml fractions. Bring fractions to a neutral pH by adding a small amount of 1.0 M Tris, pH 8.5, to each tube. Record the absorbance at 280 nm, and save the peak fractions. Combine the fractions, and dialyse them against saline.

6. Dialyse the IgG overnight in 4 litres of Coupling Buffer: 0.2 M $NaHCO_3$, 0.5 M NaCl (pH 8.5).

7. To regenerate the column for reuse, continue to wash the column with elution buffer until the absorbance at 280 nm is less than 0.10 and then wash with running buffer until the pH of the effluent is 8.9.

Protocol 5. Preparation of immunoaffinity column

1. Weigh out dry CNBr-activated Sepharose 4B (Pharmacia): 1.0 g of Sepharose per 12 mg of IgG. One gram of dry Sepharose will swell to 3.5 ml when hydrated. Couple IgG to activated Sepharose according to the following steps (13).

Protocol 5. *Continued*

2. Add dry Sepharose to 50 ml 1 mM HCl, and agitate the mixture for 15 min at room temperature.

3. Transfer the hydrated Sepharose to a 50 ml sintered glass funnel. Wash with 1 mM HCl (200 ml/g of dry gel), followed by a quick wash with Coupling Buffer.

4. Quickly transfer the Sepharose to a 50 ml plastic conical tube, and add the dialysed affinity purified IgG (*Protocol 4*). Wrap parafilm around the cap, and rotate end over end (Nutator, Clay Adams Co.) at 4 °C overnight or for 2 h at room temperature.

5. Using a 50 ml sintered glass funnel, filter off the liquid from the Sepharose. Analyse this initial effluent for protein to be sure that the protein has attached to the gel (11). Wash the gel with 500 ml of Coupling Buffer.

6. Block reactive sites left on the gel with 50 ml of 1 M ethanolamine (3.05 ml in 50 ml of Coupling Buffer). Rotate end over end (Nutator) for 2 h at room temperature. Wash with 500 ml Coupling Buffer.

7. Wash on a sintered glass filter with 250 ml each of (a), (b), (c), in that order, a total of three times:

   (a) 0.1 M acetate, 1 M NaCl, pH 4 (5.75 ml of acetic acid and 58.44 g of NaCl per litre; adjust the pH with NaOH).

   (b) 0.1 M borate, 1 M NaCl, pH 8 (6.18 g of boric acid and 58.44 g of NaCl per litre; adjust the pH with NaOH).

   (c) 0.1 M bicarbonate, 1 M NaCl, pH 9 (8.4 g of $NaHCO_3$ and 58.44 g of NaCl per litre; adjust the pH with HCl).

8. Wash with 500 ml of column buffer (5 mM phosphate, 0.15 M NaCl, 1% Trasylol (aprotinin), pH 7.4), pack the column, and store it at 4 °C.

Protocol 6. Isolation of lipoproteins by immunoaffinity column chromatography

1. Equilibrate column with column buffer (see *Protocol 5*, step 8 above).

2. Dialyse lipoprotein samples in column buffer. Cerebrospinal fluid can be applied directly. Apply sample at a very low rate (less than 5 ml/h for 5 ml of gel).

3. When all of the sample is on the column, non-specifically bound proteins are removed with 25 mM phosphate, 0.5 M NaCl, pH 7.4, until the absorbance at 280 nm returns to that of the buffer. Non-bound lipoproteins can be dialysed and retained for further analysis.

4. Elute the bound lipoproteins with 0.2 M glycine, 0.5 M NaCl, pH 2.8. The eluent from the column is collected directly into tubes containing 0.4 ml of 0.5 M Tris–HCl, pH 8.0. Dialyse the eluent into an appropriate buffer for subsequent analysis. Store gel at 4 °C in column buffer (*Protocol 5*, step 8) containing 0.01% thimerosal.

4.4 Non-denaturing gel electrophoresis

Intact lipoproteins, extracted from tissue or obtained from body fluids, can be separated by non-denaturing gel electrophoresis (14–16). The gels can be calibrated and the size of the lipoproteins determined (17) if standard lipoproteins (i.e. LDL or HDL) or proteins of known Stokes radii are separated on the same gel. The apolipoproteins present in each of the separated lipoproteins can also be identified by electrophoretically transferring the apolipoproteins to nitrocellulose paper and probing the blot with antibodies specific to particular apolipoproteins (Section 4.6).

Protocol 7. Procedure for non-denaturing gel electrophoresis

1. The separation is performed on precast 4–30% polyacrylamide gradient gels (Pharmacia or Flowgen) using a Pharmacia GE-4 electrophoresis apparatus.

2. Apply 30 μl of sample to each lane when using the 14-well comb. The sample is prepared by adding 6 μl of sample application buffer per 30 μl of sample. Prepare the application buffer by adding 20% sucrose (w/v) and bromphenol blue (1 mg/ml) as a tracking dye to the electrophoresis buffer. The electrophoresis buffer contains 10.7 g of Tris, 5 g of boric acid, and 0.93 g of EDTA per litre (pH 8.4).

3. Insert gel slabs into the upper buffer vessel of the gel apparatus, and insert the plastic sample wells into the gel until 1–2 mm of gel protrudes up into the sample wells. Place the upper vessel into the lower and connect them to the circulation pump to circulate the buffer. Remove air bubbles from the sample wells with a Pasteur pipette and prerun the gel for 15 min at 125 V before applying sample.

4. Turn the current off and slowly apply samples (30 μl containing 1–5 μg of protein/lane). Apply 5–10 μl and allow the sample to settle. Repeat until all of the sample is loaded. Turn on the circulation pump to circulate 4 °C buffer in the lower chamber only.

5. Run at 125 V for 30 min at 4 °C until the sample has entered the gel, and then circulate the buffer in both the upper and lower chambers. Electrophorese for 48 h at 125 V in a 4 °C cold room.

6. Remove the gel from the glass cassette.

4.4.1 Staining or immunoblotting

The gel can then either be stained to visualize the lipoprotein bands directly or transferred to nitrocellulose paper for detection of apolipoproteins with specific antibodies. The electrophoretic transfer to nitrocellulose is the same as described in Section 4.6 except that 0.02% SDS is added to the transfer buffer. The transfer is carried out at 4 °C for 48 h at 0.3 A. The gel is stained as follows:

(a) Incubate the gel for 30 min in 10% sulphosalicylic acid.

(b) Rinse the gel and tray with water, and stain the gel for 2 h with 0.1% Coomassie Blue R250 in 50%:5% methanol:acetic acid. Destain the gel in water containing 5%:7.5% methanol:acetic acid.

4.5 Sodium dodecyl sulphate polyacrylamide gel electrophoresis

The separation of apolipoproteins by SDS polyacrylamide gel electrophoresis (SDS–PAGE) is an extremely valuable method for determining the protein composition of tissue extracts and lipoproteins. When the total lipoprotein fraction (density < 1.21 g/ml) (Section 3.1) or individual lipoprotein fractions (Section 3.2) are separated by SDS–PAGE, an estimate of the apolipoprotein composition can be made by comparing their molecular weight to the molecular weight of standard proteins or apolipoproteins. When tissue extracts are separated on gels, the composition is too complex to determine with certainty whether apolipoproteins are present. However, if the separated proteins are electrophoretically transferred to nitrocellulose paper, the presence of specific apolipoproteins can be determined with certainty by detection with specific antibodies (5) (Section 4.6). The SDS–PAGE procedure described is a modification of the procedure of Laemmli (18, 19).

4.5.1 Preparation of samples for electrophoresis

Samples for electrophoresis should contain between 0.1 and 2 μg of each apolipoprotein of interest. Tissue extracts and isolated lipoproteins may be too dilute to analyse directly. Furthermore, samples prepared by centrifugation contain a high concentration of KBr. A high concentration of salts, both in lipoprotein density fractions and in lyophilized samples, can interfere with analysis on gels. Samples can be prepared for analysis by either of the following methods:

(a) Dialyse the sample at 4 °C into saline-EDTA or into water containing EDTA (1 mg/ml). Concentrate the sample by using a concentrator with a small membrane surface. This is conveniently done using Centricon concentrators (Amicon Corp.). Place up to 2 ml of sample in the upper chamber of the concentrator, and centrifuge it in a fixed-angle rotor until the desired volume is reached. Samples can be concentrated to as little as 50 μl using this procedure. Concentrators with molecular weight exclusions of from 3000 to 100 000 are available. The higher-molecular-weight exclusion membranes are used for lipoproteins, and the lower-molecular-weight cut-offs for apolipoproteins (the molecular weight of the smallest lipoprotein is ~5000).

(b) Alternatively, dialyse the sample into 5 mM NH_4HCO_3 and lyophilize. The NH_4HCO_3 volatilizes, leaving no residue.

If the sample contains a considerable amount of lipid (especially VLDL and chylomicrons), it should be delipidated. The lyophilized sample can be delipidated in a glass tube or a polypropylene microfuge tube as follows:

(a) Add 200 μl of 2:1 chloroform:methanol to the sample, vortex the mixture, and place it on ice for 10 min.

(b) Add 200 μl of methanol, and centrifuge the tube to pellet the protein. Aspirate the supernatant and discard. Dry the pellet under nitrogen, and solubilize it in sample application buffer.

4.5.2 Electrophoresis apparatus

Any conventional slab gel apparatus can be used for the electrophoresis, e.g. a Hoefer Series 600 gel apparatus (Hoefer Scientific). The apparatus is assembled according to the manufacturer's instructions. The gel is poured between glass plates, which are supported in a stand. The thickness of the gel can be either 0.75 mm, 1.5, or 3.0 mm, depending upon the thickness of the spacer between the plates. The number of samples can also be varied (from 10 to 20), depending upon the number of sample wells that are formed. When limited amounts of sample are available, 0.75-mm-thick gels are most appropriate. Either one or two gels can be run simultaneously in the same apparatus. Gloves should be worn to prevent contamination of the gels and plates with proteins from the skin, and for protection from non-polymerized acrylamide, *which is a neurotoxin* (20).

4.5.3 Preparation and running of the gel

There are two components to the gel: the running gel and the stacking gel (*Table 1*). The running gel is added to the plates first. After the running gel polymerizes, the stacking gel is poured, and the sample wells are formed. The first four components listed in *Table 1* can be mixed before use; however, the polymerization catalysts N,N,N',N'-tetramethylethylenediamine (TEMED) and ammonium persulphate (APS) should not be added until immediately before the gel is poured.

Table 1. Preparation of SDS polyacrylamide gels

Solution	Running Gel (percentage acrylamide acrylamide)			Stacking Gel
	5%	**10%**	**15%**	
Water (ml)	18.1	13.1	7.8	6
Buffer[a] (ml)	8.0	8.0	8.0	2.5
SDS (10%) (μl)	320	320	320	100
Acrylamide (30%)[b] (ml)	5.3	10.6	15.8	1.5
TEMED (μl)	13.4	13.4	13.4	10
Ammonium persulphate (10%) (μl)	93.4	93.4	93.4	50
Total[c] (ml)	32	32	32	10

[a] The gel buffer is 1.5 M ammediol (2-amino-2-methyl-1,3-propanediol). Ammediol (78.8 g) is adjusted to pH 8.0 with concentrated HCl, and the volume adjusted to 500 ml; the stacking buffer is 0.5 M ammediol. Ammediol (5.25 g) is adjusted to pH 6.8 with concentrated HCl, and the volume brought up to 100 ml.

[b] Because non-polymerized acrylamide is a neurotoxin, we avoid potential contact with acrylamide dust during weighing of the sample by using liquid acrylamide:bis-acrylamide (ratio 37.5:1) from American Research Products Co. The 40% solution is diluted to 30% with water and then used as described above. If it is necessary to weigh acrylamide, it is prudent to wear a particle mask and gloves and to weigh the sample in the hood to avoid the possibility of coming in contact with non-polymerized acrylamide dust.

[c] For the two 0.75-mm-thick gels, 24 ml of solution is required.

Protocol 8. Preparation of the gel

1. Mix the running gel (*Table 1*), and pour it between the glass plates until the solution reaches a level ~3 cm below the top of the plate. Carefully overlay the gel with a small amount of distilled water.

2. When the gel has polymerized (~30 min), pour off the water. Mix the stacking gel (*Table 1*), add ~1.5 ml to the polymerized lower gel, and then pour off. Add the stacking gel solution with a syringe and 20 gauge needle, and insert the comb between the glass plates to form the sample wells.

3. When the stacking gel has polymerized, remove the comb, and rinse the surface of the gel, first with deionized water and then with running buffer. Invert the apparatus onto filter paper to dry the wells.

Protocol 9. Application of sample and running gel

1. Add sample application buffer (30 μl) to the lyophilized sample (up to 20 μg of protein, but not more than 2 μg of each component), and place it in hot water (~95 °C) for 10 min before applying it to the gel. The sample application buffer contains 3% SDS, 0.76% Tris, 10% glycerol (adjusted to pH 6.8 with 1 M HCl), and bromphenol blue (1 mg/ml). Freeze the buffer in 5-ml aliquots. To reduce disulphide bonds, add 7.0 mg of dithiothreitol per ml of buffer. β-Mercaptoethanol can also be used as a reducing agent; however, it often yields an artifactual band that is seen on silver-stained gels.

2. Load the samples into the wells with a pipette, and carefully overlay them with running buffer. Concentrated running buffer (10X) is prepared as follows: 262.75 g of ammediol, 1440 g of glycine, and 100 g of SDS are brought up to a volume of 10 litres by mixing them with water. The buffer is then further diluted 1 to 10 with distilled water prior to use.

3. Place the ungreased upper chamber onto the plates and fasten them together.

4. Fill the upper chamber with ~500 ml of running buffer, pouring very slowly to avoid disturbing the samples. Make sure that the buffer covers the electrode. If there is any leakage, remove the buffer and upper chamber, find and plug the spot where the leakage occurred with Cello seal (Hoefer). Fill the lower buffer chamber with running buffer.

5. Remove the gel stand, and place the apparatus into the lower buffer chamber. Use a glass rod to remove bubbles that may be present along the bottom of the gel plates.

6. Connect the power supply to the apparatus. For 0.75 mm gels, run at 30 mA per gel, constant current, if running the gel during the day, or 5 mA per gel overnight. Keep the lower chamber cool with circulating tap water.

7. When the bromphenol blue dye marker reaches the bottom of the gels, turn the current off. Take the apparatus out of the chamber, pour off the upper buffer, and remove the plates.

8. Remove the spacers and gently pry the plates apart using an old spacer or a single-edge razor blade. The gel will stick to one plate or the other.

9. Notch the bottom left hand corner of the gel (i.e. the corner that corresponds to lane 1). The gel can either be immunoblotted (Section 4.6) or stained.

4.5.4 Coomassie Blue and silver staining

Coomassie Blue staining, as described in *Protocol 10*, is a good general staining method when adequate amounts of sample are available. The method, however, does not stain highly glycosylated proteins well and is not as sensitive as the silver-staining procedure (21) (*Protocol 11*).

Protocol 10. Coomassie Blue staining

1. Immerse the gel in stain for 1 h to overnight in a container that is covered to prevent evaporation. The stain contains 1 litre of methanol, 1 litre of water, 100 ml of acetic acid, and 2.1 g of Coomassie Blue (R-250). The solution is stirred to dissolve the components and then filtered (Whatman #1 paper).

2. Destain the gel. The destaining solution is composed of 7.5%:5% acetic acid:methanol in water. Aspirate the stain, rinse the gel with destaining solution, and then incubate the gel with destaining solution until the background on the gel clears. During destaining, gently rock the covered tray holding the gel and place one or two Kimwipes in the solution to absorb dye from the solution.

3. Store the gel in water until it is photographed or dried.

Protocol 11. Silver staining

The following procedure is based on that of Morrissey (21).

1. Put the gel in 50% methanol 10% acetic acid for 30 min and then in 5% methanol 7.5% acetic acid for 30 min.

2. Incubate the gel in 10% glutaraldehyde in water (prepare just before use) for 30 min, and then rinse it several times, each time with fresh deionized water. Leave in deionized water from 2 h to overnight, and change the water several times.

3. Incubate the gel for 30 min in water containing 5 μg of dithiothreitol/ml and then for 30 min in freshly prepared 0.1% silver nitrate. Rinse twice rapidly with water, then twice with developer: 50 μl of 37% formaldehyde in 100 ml 3% sodium carbonate (made fresh).

4. Add developer and incubate until staining reaches desired intensity.

5. Stop development by either removing developer and adding a 5% acetic acid solution, or by adding 5 ml of 2.3 M citric acid directly to the developer. Development continues for a short while after the citric acid is added.

6. Rinse the gel several times in distilled water, incubate it in 0.03% sodium carbonate for 10 min (to prevent bleaching), rinse with water, and store in water.

4.6 Immunoblotting

Intact lipoproteins separated on non-denaturing gels (14, 15) and apolipo-proteins separated by SDS–PAGE (18) can be visualized by direct staining of the gels, as already described. In addition, the proteins can be electro-phoretically transferred to nitrocellulose paper (5) and visualized by direct staining of the nitrocellulose, or the apolipoproteins can be detected and identified by using specific antibodies (5). For this purpose we routinely use a TE42 Transphor Electrophoresis Unit from Hoefer Scientific.

Protocol 12. Electrophoretic transfer

1. Soak one sponge, two sheets of filter paper that have been cut to the size of the nitrocellulose sheet, and one nitrocellulose sheet in transfer buffer. The buffer contains Tris base (75.75 g), glycine (360 g), methanol (5 litres), and isopropanol (625 ml) diluted to 25 litres with water.

2. After gel electrophoresis, remove one of the glass plates that support the gel and cut off the lower left corner of the gel to mark lane number 1. Cut the lower left corner of the nitrocellulose paper and lay it on the gel.

3. Support the nitrocellulose paper against the gel in the following manner. Place a sponge on the nitrocellulose paper (already on the gel) and the plastic holder on top of that. Turn the entire set-up over so that the glass plate is on top, and then carefully remove plate. Smooth gently over the gel with a glass rod or moist, gloved finger to remove any air bubbles between the gel and the nitrocellulose. Lay two filter papers on the gel and finally the second plastic holder. Care must be taken to ensure there are no air pockets in this 'sandwich'.

4. Fill the transfer chamber with transfer buffer (step 1, above), insert the plastic holder containing the gel, and tap several times to remove air bubbles. Because the transfer takes place from negative to positive, the holder is inserted with the gel toward the negative terminal and the nitrocellulose paper toward the positive terminal. Carry out the transfer at 4 °C for 4 h at 0.3 A or overnight at 0.15 A, with stirring.

5. After the transfer, stain the proteins on the nitrocellulose blot by soaking the blot in Ponceau S (0.2% in 3.0% trichloroacetic acid), and destain the background by rinsing the blot with distilled water. Photograph or photocopy the stained blot to facilitate lining up the autoradiogram obtained after immunoblotting. The Ponceau S stain washes off during the immunoblotting procedure. If molecular weight markers have been run, cut off the lane and stain it with Amido black (0.1% Amido black in 45:45:1 methanol:water:acetic acid, v/v) for 5 to 10 min and then destain it with 45:45:1 methanol:water:acetic acid. This staining procedure, however, shrinks the nitrocellulose slightly, and the standard strip will not

Protocol 12. *Continued*

exactly match the completed immunoblot or autoradiogram. Use Rainbow molecular weight markers (coloured standards) or Rainbow ^{14}C methylated molecular weight markers (Amersham Corporation), to visualize the molecular weight markers directly on the immunoblot or autoradiogram, respectively.

Protocol 13. Detection of protein bands with specific antibodies

1. Block non-specific binding sites in the nitrocellulose by incubation (30 min to overnight) with blotting buffer (4.44 g Tris–HCl, 2.65 g Tris base, 0.222 g $CaCl_2$, 4.68 g NaCl per litre) containing 5% non-fat dry milk (blotto buffer).

2. Decant or aspirate the buffer from the blot, and add fresh blotto buffer containing the primary antibody. The concentration of antibody or antisera used is determined empirically and differs for each antibody. Use the highest possible dilution of antibody to reduce background (1:250 or greater dilution). Incubate the blot while gently rocking the solution for 3 h at room temperature or overnight at 4 °C.

3. Briefly rinse the nitrocellulose paper twice with blotto buffer and then rinse it twice for 15 min with blotto buffer.

4. The nitrocellulose is now ready for incubation with a ^{125}I-labelled second antibody or ^{125}I-labelled Protein A. Add the iodinated second antibody or ^{125}I-labelled Protein A (0.5 to 1 × 10^6 c.p.m./ml) to the blots and incubate for 2 h at room temperature.

5. Rinse the blots twice, wash them twice for 15 min each in blotto buffer, and then rinse with water.

6. Air-dry the nitrocellulose sheet, cover it with plastic, and place it on X-ray film with an enhancing screen. Expose it for 1 to 5 days at −70 °C prior to development.

4.7 Quantitation of proteins using immunoblots

The amount of antigen present in tissue fluids or isolated lipoprotein fractions can be determined after immunoblotting by analysing the nitrocellulose paper or the autoradiogram. For this purpose, it is essential to have pure antigen available. Apply the pure protein, in a series of concentrations, to the same gel as the sample to be quantitated. Run the gel and do the immunoblot as described above (Section 4.6). Quantitate the protein by cutting the bands corresponding to the proteins of interest from the Ponceau S-stained nitrocellulose (*Protocol 12*), and count them for radioactivity. Alternatively,

the resulting autoradiogram can be used for quantitation either by densitometric scanning or by elution of the silver grains in the developed film (22).
We have found this latter procedure particularly useful because it has a
greater linear range than densitometry. If possible, enhancing screens should
not be used when the autoradiogram is to be used for quantitation because
the intensity may not be proportional to the concentration. It is essential that
standards of known concentrations are present on the autoradiogram
containing the samples.

Protocol 14. Quantitation of immunoblot autoradiogram by elution of
silver grains

1. Wipe the autoradiogram with an antistatic cloth to remove dust.
2. Cut out the bands corresponding to the standards and to the protein to be
   quantified from the autoradiogram. Cut each of these bands into small
   pieces, and put them in a glass tube. As blanks, process several pieces of
   the same film that have not been exposed to radioactivity; these pieces
   should be similar in size to the sample bands.
3. Add 1 ml of 1 M NaOH to each sample, and incubate them at room
   temperature for from 2 h to overnight with occasional vortexing.
   Continue incubation until all of the silver grains are eluted from the film.
4. Add glycerol (0.4 ml) to the sample, vortex, and record the absorbance of
   the supernatant at 500 nm. A standard curve is generated based on the
   standards that were run on the same gel, and the concentration of the
   unknown protein is determined by reference to the standard curve.

4.8 Neuraminidase hydrolysis of apolipoprotein E

Apolipoprotein E on SDS–PAGE gels often appears as a series of bands that
vary in apparent molecular weight (M_r ~35 000–37 500). This heterogeneity
is often due to variable levels of sialylation (23). To determine whether the
heterogeneity is due to sialylation, proteins are analysed before and after
neuraminidase digestion by SDS–PAGE or by two-dimensional gel electrophoresis (1, 24).

The enzyme used is from *Clostridium perfringens* (Sigma N-5631, Sigma
Chemical Co.). Because of the instability of the enzyme, it is stored at −20 °C
in its unopened foil pouch until use (within 3 months of arrival). One
complete vial is used for an assay, and the enzyme should not be stored again
once it has been opened. To ensure that the digestion is complete, a high
enzyme-to-protein ratio is used. In addition, a second addition of enzyme is
made half-way through the incubation period. While pH 5 is optimum for
enzymatic activity, the hydrolysis of apo E is carried out at pH 4.0 because
apo E precipitates at pH 5.

Protocol 15. Neuraminidase treatment of apolipoprotein E

1. Dialyse the lipoproteins or apolipoproteins into 5 mM NH_4HCO_3. Aliquot samples containing 1 to 10 µg of the protein of interest into duplicate 13 × 100 mm disposable glass tubes and lyophilize them.
2. Prepare the digestion buffer (0.1 M ammonium acetate, pH 4.0) by adjusting the pH of 0.1 M acetic acid (5.75 ml of glacial acetic acid/liter) to pH 4.0 with concentrated NH_4OH (~2.0 ml).
3. Dissolve lyophilized samples in 100 µl of digestion buffer.
4. Dissolve one vial of enzyme (0.63 mg) in 1.26 ml of digestion buffer by swirling, then use it immediately.
5. Add 45 µl of enzyme solution to test samples and 45 µl of enzyme buffer to duplicate control samples. Cover the tubes with parafilm and incubate them for 2 h at 37 °C; gently agitate the contents once or twice during incubation. After 2 h, repeat the additions of buffer with or without enzyme and incubate the samples for another 2 h at 37 °C.
6. Lyophilize the samples and delipidate them prior to gel electrophoresis if intact lipoproteins were hydrolysed.

5. Lipid analysis

5.1 Extraction and analysis of lipids

Extraction of lipids is conducted in solvent-washed (2:1 chloroform:methanol, v/v) glass tubes. *Protocol 16* is a modification of the procedure described by Kates (25). The sample is then analysed as detailed in *Protocol 17.*

Protocol 16. Procedure for the extraction of lipids

1. For 1 ml of sample, add 1.25 ml of chloroform and 2.5 ml of methanol to form a one-phase system. The proportions are critical; adjust the solvent volumes proportionately to the sample size. Vortex, leave on ice for 15 min, and centrifuge for 5 min at 1700*g* to pellet the protein.
2. Remove the supernatant to a clean tube and add 1.25 ml of $CHCl_3$ and 1.25 ml of water containing 0.88% KCl. Vortex and centrifuge for 5 min at 1700*g* to separate layers. The bottom layer contains most of the lipids. The upper aqueous layer may contain some gangliosides.
3. Carefully transfer the solvent from the bottom of the tube to a clean, solvent-rinsed tube, and dry it under a stream of nitrogen in water bath at 37 °C. Dissolve the sample in a known volume of choroform:methanol

(2:1), and store it at $-70\,°C$ until analysed. Butylated hydroxytoluene (0.05%) may be added to the chloroform:methanol to prevent oxidation of the sample.

Protocol 17. Analysis of phospholipid phosphorus, total cholesterol, and triacylglycerols

1. Phosphorus analysis:[a]

(a) Acid-wash glass tubes (13 × 100 mm) with 0.2 M HCl in a large beaker. Bring to a boil, cool to room temperature, and rinse the tubes at least three times with deionized water prior to use.

(b) Pipette standards (0.05–3.2 µg of phosphorus (Sigma #661–9, 20 µg/ml)) and samples in triplicate. The volume is not important since the water evaporates.

(c) Add 300 µl of 5 M H_2SO_4 to samples and standards, and heat them at 180 °C for 12 to 16 h.

(d) Cool the tubes, and add 100 µl of 30% H_2O_2. Place the tubes back in the 180 °C heating block for a minimum of 3 h. The samples should be clear at this point. If the samples are not clear, add more 30% H_2O_2, and heat for another 3 h.

(e) Cool the samples, and add 650 µl of water and 200 µl of 4% ammonium molybdate (BANCO, Anderson Laboratories, Inc.), that has been filtered through #4 Whatman filter paper, and vortex well.

(f) Add 50 µl of Fiske & Subbarow reducer (Sigma) to each tube and vortex. The reagent is prepared by dissolving 1.5 g of reducer in 9.45 ml of deionized water.

(g) Heat the solution in the heating block at 100–110 °C for 8 min, and then determine the absorbance at 830 nm.

(h) Calculate unknowns using linear regression of the phosphorus standards. Convert the phosphorus value to that of phospholipid by multiplying by 25.

2. Cholesterol analysis:

(a) Dry the lipid extract containing 2 to 20 µg of cholesterol under a stream of nitrogen in solvent-rinsed tubes. Pipette cholesterol standards (1 to 25 µg) into glass tubes and evaporate to dryness (Sigma cholesterol standard, 5 mg/ml in isopropanol).

(b) Add 250 µl of Lieberman/Burchard reagent (Sigma), and incubate for 15 min at 37 °C Vortex immediately after adding reagent and again before reading the absorbance at 630 nm.

233

Protocol 17. *Continued*

3. Triacylglycerol analysis:

(a) Evaporate the lipid extract containing 1 to 30 µg of triacylglycerol to dryness. Pipette triolein standards (0.6 to 35 µg) into glass tubes that have been solvent-rinsed and evaporate to dryness (Sigma triolein standard, 3 mg/ml in isopropanol).

(b) Incubate with 300 µl of BMD enzymatic triacylglycerides GPO reagent (Boehringer–Mannheim) for 15 min at 37 °C; vortex the sample immediately after adding reagent and before reading absorbance at 500 nm.

a The phosphorus analysis is a modification of the procedure of Bartlett (26).

5.2 Thin-layer chromatography

5.2.1 Qualitative analysis

We have found the procedure in *Protocol 18* useful for the separation and qualitative estimation of the amounts of phospholipids and neutral lipids on the same high-performance (HP) TLC plate (27).

Protocol 18. Qualitative analysis of phospholipids and neutral lipids

1. High-performance TLC is performed on 10 × 20 cm plates (HPTLC LHP-K plates) with a preadsorbent area (Whatman).

2. With a pencil, mark the plate 4.5 cm and 7.0 cm above the top of the preadsorbed area.

3. Develop the plate twice in methanol to move contaminants to the top of the plate, and then develop the plate in methyl acetate:1-propanol: choloroform:methanol:0.25% KCl (25:25:25:10:0.9 by volume).

4. Dry the plate for 15 min at 100 °C. Cool the plate, and apply the sample 0.75–1.0 cm below the top of preadsorbed area.

5. Develop the TLC plate to the 4.5 cm mark in the solvent system described above in step 3. Following development, dry with hot air from a hair dryer for 10 min.

6. Develop the plate to the 7 cm line in a system of 75:23:2 hexane:diethyl ether:acetic acid. Dry with hot air for 10 min.

7. Develop the plate to the top (10 cm) with hexane.

8. Visualize the lipids by charring the plates. Dip the plates into 10% (w/v) cupric sulphate in 8% (w/v) phosphoric acid and char them at 180 °C until all bands are visible (28).

5.2.2 Quantitative analysis

When HPTLC is to be used for quantitation of lipids, extra precautions need to be followed. A low background level on the TLC plate following charring is essential. Filter the solvents through a 0.2 μm PTFE (polytetrafluoro-ethylene) membrane filter (Gelman Sciences). This removes dust and lint from the solvents, substantially reducing the background charring (29).

The application and development of the sample as tight bands facilitate the detection of low levels of lipid. This is facilitated by using a Camag Nanomat spotter with a 0.5 μl pipette tip. Sample can be applied up to three times for a total volume of 1.5 μl. The use of a Camag linear developing chamber in combination with the spotter improves the quantitation of neutral lipids (30), since the neutral lipids develop as much more dense, tight bands than when developed in a conventional TLC tank. When this system is used, and the samples are applied to the plate at 0.5 cm intervals, 39 samples can be spotted on each side of an HPTLC plate, and the samples are simultaneously developed from both sides toward the centre (*Figure 1*). Standards are applied to the same HPTLC plate as the samples and used to prepare a standard curve. The range of values being quantified must be in the linear range of the detection system.

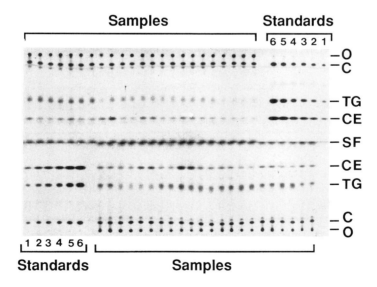

Figure 1. Use of the Camag linear developing chamber to develop neutral lipids. Lipid standards and samples were applied (origin, O) on both sides of the HPTLC plate and were simultaneously developed from both sides until the solvent fronts (SF) met at the centre of the plate. The standards contain equal amounts of cholesterol (C), triacylglycerol (TG), and cholesteryl ester (CE). The mass of each component in lanes 1, 2, 3, 4, 5, and 6 is 0.1, 0.2, 0.3, 0.5, 1.0 and 1.5 μg, respectively. The lipids were visualized by charring.

The lipids can be quantified only if the composition is sufficiently simple to allow for adequate separation of all components. If this is not possible using a single TLC plate for neutral and polar lipids as described above (*Protocol 18*), individual systems for the separation may be required.

Protocol 19. Quantitative analysis of phospholipids and neutral lipids

1. Immerse the HPTLC plates (Silica gel 60, 10 × 20 cm, E. Merck), in methanol for 1 h to remove the contaminants, air-dry them, and activate them for 1 h at 110 °C.

2. Using the Camag Nanomat spotter with a 0.5 µl pipette, apply standards (0.1–1.5 µg) and samples in a total volume of 1.5 µl or less.

3. For neutral lipids, develop the plates in the Camag linear developing chamber in a system of 63:18.5:18.5:1 hexane:heptane:diethyl ether:acetic acid. Remove the plate after the two solvent fronts meet at the center of the plate (30) (*Figure 1*).

4. For polar lipids, develop the plates in a standard TLC developing chamber with *n*-propanol:methyl acetate:chloroform:methanol:43 mM KCl (25:25:25:10:9) (27, 30). Remove the plates when the solvent has migrated 9 cm.

5. After air-drying, dip the plate in 10% cupric sulphate in 8% phosphoric acid for 20 sec. Place the plate in a cool oven, heat to 180 °C, and turn the oven off. When the plate is cool, thoroughly clean the back of the plate and scan it using a densitometer in the reflectance mode.

---

6. Immunocytochemistry

Light-level immunocytochemistry is used to determine which cells in tissues contain apolipoproteins and to detect other antigens, such as lipoprotein receptors (3). For immunocytochemistry to be successful, excellent antibodies must be available. We have found it necessary to prescreen the serum from animals for cross-reactivity before using them for antibody production. For this purpose, serum from candidate animals is used at 1:2000 to 1:10 000 dilution, and the immunocytochemistry procedure described below is performed on the tissues of interest to ensure that only animals that give low backgrounds are used for antibody production. Specific pathogen-free rabbits appear to have lower levels of cross-reactive antibodies than normal rabbits. Preimmune serum from the animals selected for antibody production is saved and is used as a negative control. It is possible to obtain erroneous results from immunocytochemistry experiments because of cross-reactivity of antisera with antigens in specific tissues or within specific cells types. This can occur even though the antisera appear to be monospecific when tested against plasma. The monospecificity of the antibody should be tested on extracts of the tissue of interest to ensure that only one antigen of the appropriate molecular weight is detected (*Protocol 13*). The immunocytochemistry procedure is essentially that described by Boyles *et al.* (3).

Protocol 20. Preparation of 4% paraformaldehyde

1. Dissolve 4.0 g of paraformaldehyde (Sigma) in 40 ml of distilled water by heating on a hot plate. When condensation appears on the neck of the flask, the paraformaldehyde is solubilized by dropwise addition of 1 M NaOH, until the solution clears.

2. Cool the solution on ice to room temperature, and adjust the pH to 7.0 with 1 M HCl.

3. Add 50 ml of 0.3 M phosphate buffer, pH 7.4 (0.3 M Na_2HPO_4 adjusted to pH 7.4 with 0.3 M NaH_2PO_4). Bring the volume to 100 ml with deionized water, and recheck the pH.

Protocol 21. Preparation of gelatin-coated microscope slides

1. Immerse slides in running water for about 15 min and then rinse in double-distilled water.

2. Clean slides in 95% ethanol for 10 min and drain.

3. Dip the slides into gelatin solution at room temperature. The gelatin solution is prepared as follows: Dissolve gelatin (6.0 g) in 288 ml of double-distilled water by heating to 56 °C, then add 120 ml of 95% ethanol, 20 ml of 2% chromium potassium sulphate, and 28 ml of glacial acetic acid. Drain the slides well, air-dry them, and bake them for exactly 1 h at 60 °C.

6.1 Preparation of tissue

To reduce background reactivity from antigens in the blood, animals are routinely perfused with ice-cold PBS as described (Section 2), at a constant pressure of 110 mm Hg until the fluid flowing from the vena cava is free of blood and the liver is well cleared. Tissue can be obtained at this point, or the whole animal can be perfusion-fixed using freshly prepared 4% para-formaldehyde (*Protocol 20*).

Protocol 22. Preparation of tissue for immunocytochemistry

1. Remove tissues from the animal and prepare them for freezing and storage. Small pieces of the tissue can be cryoprotected prior to freezing by incubation at 4 °C for 12–16 h in PBS containing 18% sucrose, or they can be frozen directly.

2. Cut tissue with a sharp razor blade into pieces (~2 to 3 mm in largest dimension), and drop them into a small beaker containing 2-methyl butane pre-cooled with liquid nitrogen. This snap-freezing procedure prevents the formation of large ice crystals, which can disrupt the cells. Rapid freezing is essential: If tissue pieces are large, the surface of the tissue may be frozen rapidly enough to obtain satisfactory results, but the internal portion of the tissue may be damaged as a result of the formation of ice crystals.

3. Store the frozen tissue in liquid nitrogen until used.

6.2 Immunocytochemistry

The tissue is mounted onto the block holder for sectioning using OCT (Miles Scientific), as the embedding medium. The tissue remains frozen during embedding. Coat the block holder with OCT, and place it on the cryobar. As the OCT starts to solidify, the frozen tissue is placed on it and immediately sprayed with Accu-Freeze (Stephens Scientific) to enhance the freezing. Allow time for the tissue block to equilibrate to the cryostat temperature (between −15 °C and −30 °C) before cutting the sections. The exact temperature depends on the tissue. Cut frozen sections (5 to 10 μm), pick them up on gelatin-coated slides (*Protocol 21*), and air-dry them for 1 h. The tissue can them be briefly fixed in 4% paraformaldehyde (*Protocol 20*) if it was not fixed initially. To facilitate the use of small volumes of reagents (200–500 μl per slide) during immunocytochemistry, form a circular dam around the tissue sections using a PAP pen (Kiyota Express). Conduct incubations with the slides in a plexiglass incubation (humidity) chamber (Accurate Chemical & Scientific Corp.), to prevent evaporation. If the antigen is intracellular, include Triton X-100 in the dilution buffer to permeabilize the cells. Omit the detergent if the antigen is on the cell surface. Initially, a range of antisera dilutions (1:500 to a 1:20 000) are used to determine the optimal concentration. To conserve reagent and to ensure specificity, the lowest level of antibody possible should be used. All antisera contain contaminating antibodies, and if too high a concentration of antisera is used, minor antibodies that cross-react with antigens in the tissue will raise the background. As controls, run preimmune sera and slides in which the primary antibody is omitted.

The immunocytochemistry procedure is extremely sensitive. The tissue sections are incubated with primary antibody, with a biotinylated second antibody, and then with a biotin–avidin horseradish peroxidase complex (ABC reagent) before development of reaction product. These reagents (except for the primary antisera) can be purchased in a kit from Vector Laboratories.

Protocol 23. Immunocytochemistry procedure

1. Incubate tissue sections for 30 min to 3 h at 4 °C with a buffer that is intended to quench (block) non-specific binding sites. The quench buffer contains 0.1 g of non-fat dry milk, 1.16 g ammonium acetate, 0.48 g of Dulbecco's PBS (Gibco Laboratories), and 2 ml of blocking serum per 100 ml. Blocking serum is serum from the species in which the biotinylated antibody was produced. Add Triton X-100 (0.15%) if desired. Remove excess quench buffer from tissue sections by tipping slides onto paper towels (do not wash slides before applying primary antibody).

2. Dilute the antisera in PBS (9.6 g/litre) containing 0.1% non-fat dry milk, 0.015 M ammonium acetate, 0.1% Triton X-100, 0.2% blocking serum, and incubate with the tissue for 16 h at 4 °C. This buffer, but without blocking serum and Triton (referred to as wash buffer), is used in subsequent washes and in making up the secondary antibody and the ABC reagent.

3. Remove this solution from the tissue sections, and place the slides in a staining dish with wash buffer for 8–16 h at 4 °C.

4. Incubate the tissue sections for 3 h at room temperature with the biotinylated secondary antibody (40 µl per 10 ml of wash buffer). In some cases the second antibody may react with IgG in the tissue and give a false positive. This will be apparent when the controls without primary antibody are examined. If this occurs, the concentration of the secondary antibody may be reduced and the secondary antibody can be adsorbed using 10% normal serum from the same species as the tissue samples. (This is done by incubating the diluted secondary antibody with the normal serum for 16 h at 4 °C before applying it onto tissue sections.)

5. Wash the tissue sections for 2 h at room temperature in a staining dish containing the wash buffer.

6. To inhibit endogenous peroxidase activity, incubate the tissue sections with 0.3% hydrogen peroxide in methanol for 30 min at room temperature.

7. Rinse the tissue sections thoroughly with wash buffer.

Protocol 23. *Continued*

  8. The ABC reagent (prepared at the same time as the secondary antibody to allow time for the avidin–biotin complex to form) is incubated with the tissue sections for 2–3 h at room temperature. The ABC reagent is prepared in the wash buffer as described in the package insert.

  9. Wash the tissue sections for 2–16 h at 4 °C with the wash buffer.

  10. Incubate the slides for 15–30 min at room temperature in 0.1 M monobasic phosphate buffer, (adjust to pH 6.5 with ammonium hydroxide), containing 2.5 mM nickel chloride and 0.2 M diamino-benzidine (DAB). Nanopure water should be used in making up the substrate solution, which should be prepared just prior to use. Note that DAB is a suspected carcinogen.

  11. To develop the peroxidase reaction, add 0.03% hydrogen peroxide to the substrate solution, allow to react for 2 min and then wash the tissue sections twice in PBS and twice with distilled water to remove salts. The reaction results in a black precipitate at the site of antigen–antibody complex formation.

  12. Counterstain for 5 min with 1% methyl green, rinse in distilled water, and dehydrate by dipping twice in 95% ethanol, twice in 100% ethanol, and twice in xylene, and then mount in Cytoseal (Stephens Scientific).

7. Functional characterization of tissue lipoproteins

If tissue lipoproteins can be obtained in large quantities, any of the standard techniques for the assessment of function could be applied. Unfortunately, this is usually not the case. The strategy for determination of function must therefore be carefully thought out to allow for the use of the minimal amount of sample. An example of a procedure used to demonstrate that the apo E-containing lipoproteins in cerebrospinal fluid (CSF) are bound and internalized by LDL receptors is presented.

7.1 Competitive binding studies

In the characterization of the apo E-containing lipoproteins in canine CSF, we were able to prove that they bind to the LDL receptor by using a competitive binding assay (1). This procedure is described in detail in Chapter 6. For this study the apo E content of the CSF lipoproteins was determined by analysis of autoradiograms of Western blots of the lipoproteins as described above (Section 4.7 and *Protocol 14*). To carry out the assay, we made use of the fact that both apo B-containing lipoproteins, such as LDL, and apo E-containing lipoproteins bind to the LDL receptor and that apo E-containing lipoproteins bind with a much higher affinity than do LDL. For this reason, we tested the ability of the CSF apo E-containing lipoproteins to compete for the binding of ^{125}I-labelled LDL. Before the competitive binding experiment, human fibroblasts were grown in 35 mm dishes under conditions that up-regulate the LDL receptor. The cells were then incubated at 4 °C with ^{125}I-labelled LDL (2 μg/ml) alone or with increasing concentrations of CSF lipoproteins containing apo E. Ninety percent competition was obtained when CSF lipoproteins containing 1 μg of apo E/ml were added. This experiment could therefore be performed in duplicate using CSF lipoproteins containing only 4 μg of apo E.

Acknowledgements

We thank Kerry Humphrey and Joan Ketchmark for manuscript preparation, Al Averbach and Sally Gullatt Seehafer for editorial assistance and Jim McGuire, Annabelle Friera, and Kay Arnold for critical review of the manuscript. This work was supported in part by National Institutes of Health grants NS 25678 and HL 41633.

References

1. Pitas, R. E., Boyles, J. K., Lee, S. H., Hui, D. Y., and Weisgraber, K. H. (1987). *J. Biol. Chem.*, **262**, 14352.
2. Boyles, J. K., Pitas, R. E., Wilson, E., Mahley, R. W., and Taylor, J. M. (1985). *J. Clin. Invest.*, **76**, 1501.
3. Boyles, J. K., Zoellner, C. D., Anderson, L. J., Kosik, L. M., Pitas, R. E., Weisgraber, K. H., Hui, D. Y., Mahley, R. W., Gebicke-Haerter, P. J., Ignatius, M. J., and Schooter, E. M. (1989). *J. Clin. Invest.*, **83**, 1015.
4. Hoff, H. F. and Morton, R. E. (1985). *Ann. N.Y. Acad. Sci.*, **454**, 183.
5. Towbin, H., Staehelin, T., and Gordon, J. (1979). *Proc. Natl. Acad. Sci. USA*, **76**, 4350.
6. Havel, R. J, Eder, H. A., and Bragdon, J. H. (1955). *J. Clin. Invest.*, **34**, 1345.
7. Redgrave, T. G., Roberts, D. C. K., and West, C. E. (1975). *Anal. Biochem.*, **65**, 42.
8. Hatch, F. T. and Lees, R. S. (1968). *Adv. Lipid Res.*, **6**, 1.
9. Holmquist, L. (1982). *J. Lipid Res.*, **23**, 1249.
10. Weisgraber, K. H. and Mahley, R. W. (1980). *J. Lipid Res.*, **21**, 316.
11. Lowry, O. H., Rosebrough, N. J., Farr, A. L., and Randall, R. J. (1951). *J. Biol. Chem.*, **193**, 265.
12. Gibson, J. C., Rubinstein, A., Ngai, N., Ginsberg, H. N., Le, N. A., Gordon, R. E., Goldberg, I. J., and Brown, W. V. (1985). *Biochim. Biophys. Acta*, **835**, 113.
13. Pharmacia Laboratory Separation Division (1990). *Separation News*, **13**, 1.
14. Krauss, R. M. and Burke, D. J. (1982). *J. Lipid Res.*, **23**, 97.
15. Nichols, A. V., Gong, E. L., and Blanche, P. J. (1981). *Biochem. Biophys. Res. Commun.*, **100**, 391.

16. Gordon, V., Innerarity, T. L., and Mahley, R. W. (1983). *J. Biol. Chem.*, **258**, 6202.
17. Blanche, P. J., Gong, E. L., Forte, T. M., and Nichols, A. V. (1981). *Biochim. Biophys. Acta*, **665**, 408.
18. Laemmli, U. K. (1970). *Nature*, **227**, 680.
19. Wetterau, J. R., Aggerbeck, L. P., Rall Jr, S. C., and Weisgraber, K. H. (1988). *J. Biol. Chem.*, **263**, 6240.
20. Johnstone, A. and Thorpe, R. (1989). *Immunochemistry in Practice*. Blackwell Scientific Publications, Oxford.
21. Morrissey, J. H. (1981). *Anal. Biochem.*, **117**, 307.
22. Suissa, M. (1983). *Anal. Biochem.*, **133**, 511.
23. Zannis, V. I., Breslow, J. L., Utermann, G., Mahley, R. W., Weisgraber, K. H., Havel, R. J., Goldstein, J. L., Brown, M. S., Schonfeld, G., Hazzard, W. R., and Blum, C. (1982). *J. Lipid Res.*, **23**, 911.
24. Weisgraber, K. H., Rall Jr, S. C., Innerarity, T. L., Mahley, R. W., Kuusi, T., and Ehnholm, C. (1984). *J. Clin. Invest.*, **73**, 1024.
25. Kates, M. (1972). In *Laboratory Techniques in Biochemistry and Molecular Biology. Part II. Techniques of Lipidology. Isolation, Analysis, and Identification of Lipids* (ed. T. S. Work and E. Work), Vol. 3, North-Holland/American Elsevier, Amsterdam.
26. Bartlett, G. R. (1959). *J. Biol. Chem.*, **234**, 466.
27. Yao, J. K. and Rastetter, G. M. (1985). *Anal. Biochem.*, **150**, 111.
28. Bitman, J. and Wood, D. L., (1982). *J. Liquid Chromatogr.*, **5**, 1155.
29. Wood, W. G., Cornwell, M., and Williamson, L. S. (1989). *J. Lipid Res.*, **30**, 775.
30. Schmitz, G., Assmann, G., and Bowyer, D. E. (1984). *J. Chromatogr.*, **307**, 65.

A1

Suppliers

Accurate Chemical and Scientific Corp., 300 Shames Drive, Westbury, NY 11590, USA.

Alma-Research, 50142 Firenze, Italy.

American Research Products: Chemlab Ltd, Nuffield Road, Cambridge, CB4 1TH, UK. **American Research Products Co.**, 30175 Solon Industrial Parkway, Solon, Ohio 44139, USA.

American Type Culture Collection, 12301 Parklawn Drive, Rockville, MD 20852–1776, USA

Amersham: Amersham International PLC, Lincoln Place, Green End, Aylesbury, Bucks HP20 2TP, UK. **Amersham Corporation**, 2636 South Clearbrook, Arlington Heights, IL 60005, USA.

Amicon: Amicon Ltd, Upper Mill, Stonehouse, Gloucestershire, GL10 2EJ, UK. **W.R. Grace & Co., Amicon Division**, 24 Cherry Hill Drive, Danvers, MA 01923, USA.

Analytichem: Varian Sample Preparation Products, European Technical Centre, PO Box 234, Cambridge CB2 1PE, UK. **Varian Sample Preparation Products, Inc.**, 24201 Frampton Avenue, Harbor City, CA 90710, USA.

Anderson: Anderson Laboratories, Inc., 5901 Fitzhugh Avenue, Fort Worth, TX 76119, USA.

Anton Paar: Paar Scientific Ltd, 594 Kingston Road, Rayner Park, London SW20, UK. **Anton Paar USA**, 1030A Wilmer Avenue, Richmond, VA 23227, USA.

Avanti Polar Lipids, 700 Industrial Park Road, Alabaster, Alabama 35007, USA.

Baker: Linton Instrumentation, Hysol, Harlow CM18 6QZ, UK. **J.T. Baker Chemical Co.**, Phillipsburg, NJ 08865, USA.

BDH: Merck Ltd, Broom Road, Poole, Dorset BH12 4NN, UK. **Gallard Schlesinger Industries Inc.**, 584 Mineloa Avenue, Carle Place, NY 11514–1731, USA.

Beckman: Beckman Instruments (UK) Ltd, Progress Road, Sands Industrial Estate, High Wycombe, Bucks HP12 4JL, UK. **Beckman Instruments Inc.**, 2500 Harbor Road, Fullerton, CA 92634–3100, USA.

Behring Diagnostics: Hoechst UK Ltd, Hoechst House, 50 Salisbury Road, Hounslow, Middx TW4 6JH, UK. **Behring Diagnostics, Inc.**, 17 Chubb Way, Summerville, NJ 08876, USA.

Biomakor, Kiryat Weimann, 76326 Rehovot, Israel.

Biomeda Corporation, PO Box 8045, 1155 – E Triton Drive, Foster City, CA 94404, USA.

Biometra Ltd, PO Box 167, Maidstone, Kent ME14 2AT, UK.

Bio-Rad: Bio-Rad Laboratories, Ltd, Bio-Rad House, Maylands Avenue, Hemel Hempstead, Herts HP2 7TD, UK. **Bio-Rad Laboratories**, 3300 Regatta Blvd, Richmond, CA 94804, USA.

Biosoft: Biosoft, 22 Hills Road, Cambridge CB2 1JP, UK. **Biosoft**, PO Box 10938, Ferguson, MO 63135, USA.

Biostat Diagnostics Ltd, Biostat House, Pepper Road, Hazel Grove, Stockport, Cheshire SK7 5BW, UK.

Boehringer: Boehringer Mannheim UK (Diagnostics and Biochemicals) Ltd, Bell Lane, Lewes, East Sussex BN17 1LG, UK. **Boehringer Mannheim Diagnostics**, PO Box 50414, Indianapolis IN 46250, USA.

Branson: Lucas Dawe Ultrasonics Ltd, Concord Road, Western Avenue, London W3 0SD, UK. **Branson Ultrasonics**, Danbury, CT 06813, USA.

Brinkman: Chemlab Scientific Products, Construction House, Grenfel Avenue, Hornchurch, Essex RM12 4EH, UK. **Brinkman Instruments**, 1 Cantiague Road, PO Box 1019, Westbury, NY 11590, USA.

Buchler: Arnold R. Horwell Ltd, 73 Maygrove Road, West Hampstead, London NW6 2BP, UK. **Buchler Instruments Inc.**, 8811 Prospect, Kansas City, Missouri 64132, USA.

Calbiochem: Novabiochem (UK) Ltd, 3 Heathcoat Building, Highfields Science Park, University Boulevard, Nottingham NG7 2QJ, UK. **Calbiochem Corp.**, PO Box 12087, San Diego, CA 92112–4180, USA.

Camag: Baird and Tatlock Ltd, PO Box 1, Romford, Essex RM1 1HA, UK. **Camag**, Wilmington, NC 28405, USA.

Cedarlane: Cedarlane Laboratories, 5516–8th Line, R.R.2, Hornby, Ont. LOP 1EO, Canada. **Vector Laboratories**, 16 Wulfric Square, Bretton, Peterborough PE3 8RF, UK. **Accurate Chemical and Scientific Corporation**, 300 Shames Drive, Westbury, NY 11590, USA.

Centers for Disease Control, Department of Health and Human Services, Atlanta, Georgia 30333, USA.

Chrompack: Chrompack UK Ltd, Unit 4, Indescon Court, Millharbour, London E14 9TN, UK. **Chrompack Inc.**, PO Box 6795, Bridgewater, NJ 08807–0795, USA.

Clay Adams: Arnold R. Horwell Ltd, 73 Maygrove Road, West Hampstead, London NW6 2BP, UK. **Clay Adams Co.**, Division of Becton Dickinson Primary Care Diagnostics, PO Box 370, Sparks, MD 21152, USA.

Coulter: Coulter Electronics Ltd, Northwell Drive, Luton, Beds LU3 3RH, UK. **Coulter Corporation**, PO Box 2145, Hialeah, FL 33012–0145 USA.

DAKO: DAKO Ltd, 16 Manor Courtyard, Hughenden Avenue, High Wycombe, Bucks HP13 5RE, UK. **DAKO Corporation**, 6392 Via Real, Carpinteria, CA 93013, USA.

DuPont: DuPont (UK) Ltd, Diagnostics and Biotechnology Division, Wedgwood Way, Stevenage, Herts SG1 4QN, UK. **E.I. DuPont de Nemours and Co. (Inc.)**, Photo Products Dept, Wilmington, DE 19898, USA.

Dynatech: Dynatech Laboratories Ltd, Daux Road, Billinghurst, West Sussex RH14 9SJ, UK. **Dynatech Laboratories Inc.**, Biotechnology Products, 14340 Sullyfield Circle, Chantilly, VA 22021, USA.

The Enzyme Center, 36 Franklin Street, Malden, MA 02148, USA.

Europa: Europa Scientific Ltd, Scope House, Weston Road, Crewe, Cheshire CW1 1DD, UK. **Metabolic Solutions, Inc.**, 33 Nagog Park, Acton, MA 01720, USA.

Falcon Labware: Becton Dickinson UK Ltd, Between Towns Road, Cowley, Oxford OX4 3BR, UK. **Becton Dickinson**, 2 Bridgewater Lane, Lincoln Park, NJ 07035, USA.

Flow: ICN Flow Ltd, Eagle House, Peregrine Business Park, Gomm Road, High Wycombe, Bucks HP13 7DL, UK. **Flow Laboratories Inc.**, 7655 Old Springhouse Rd, McLean, VA 22102, USA.

Flowgen Instruments Ltd., Broad Oak Enterprise Village, Broad Oak Road, Sittingbourne, Kent ME9 8AQ, UK.

Gelman: Gelman Sciences Ltd, 10 Harrowden Road, Brackmills, Northampton NN4 0EZ, UK. **Gelman Sciences**, Ann Arbor, MI 48106, USA.

Gibco: Life Technologies Ltd, PO Box 35, Trident House, Renfrew Road, Paisley PA3 4EF, UK. **Life Technologies, Inc.**, 8717 Grovemont Circle, Gaithersburg, MD 20877, USA.

Hoefer: Hoefer Scientific Instruments UK, Unit 12, Croft Road Workshops, Off Hempstalls Lane, Newcastle under Lyme, Staffs ST5 0TW, UK. **Hoefer Scientific Instruments**, 654 Minnesota Street, San Francisco, CA 94107, USA.

Ilford: Ilford Photo Co., 14–22 Tottenham Street, London W1P 0AH, UK. **Ilford Photo Corp.**, West 70 Century Road, PO Box 288, Paramus, NJ 07653, USA.

Isco: Jones Chromatography, New Road, Hengoed, Mid Glamorgan, S. Wales CF8 8AU, UK. **Isco, Inc.**, PO Box 5347, Lincoln, NE 68505, USA.

Jackson Immunoresearch Laboratories: Stratech Scientific Ltd, 61–63 Dudley Street, Luton, Beds LU2 0NP, UK. **Jackson Immunoresearch Laboratories Inc.**, 872 West Baltimore Pike, PO Box 9, West Grove, PA 19390, USA.

Jandel: The Core Store Ltd, The Studio, Hawthorn Cottage, Marbury Road, Comberbach, Cheshire CW9 6AU, UK. **Jandel Scientific**, Corte Madera, CA 94925, USA.

Joyce-Loebl: Joyce-Loebl Ltd, Dukesway, Team Valley, Gateshead, Tyne and Wear NE11 0PZ, UK. **JL Inc.**, 1250 Oakmead Parkway, Suite 210, Sunnyvale, CA 94087, USA.

Kabi: Kabi Pharmacia, Ltd, Davy Avenue, Knowlhill, Milton Keynes MK5 8PH, UK. **Kabi Pharmacia**, 800 Centennial Ave, PO Box 1327, Piscataway, NJ 08855–1327, USA.

Kinematica: Philip Harris Scientific, 618 Western Avenue, Park Royal, London W3 0TE, UK. **Brinkman Instruments**, Westbury, NY 11590, USA.
Kiyota Express, 1940 East Devon Avenue, Elkgrove Village, IL 60007, USA.
Kodak: Phase Separations Ltd, Deeside Industrial Park, Deeside, Clwyd CH5 2NU, UK. **Eastman Fine Chemicals**, Eastman Kodak Company, LRPD-1001 Lee Road, PO Box 92822, Rochester, NY 14692–7073, USA.
Lab Tek: see Nunc
LKB: see Pharmacia
Merck: Merck Ltd, Broom Road, Poole, Dorset BH12 4NN, UK. **EM Science**, 111 Woodcrest Road, Cherry Hill, NJ 08034–0395, USA
Miles: ICN Flow Ltd, Eagle House, Peregrine Business Park, Gomm Road, High Wycombe, Bucks HP13 7DL, UK. **Miles Incorporated**, 1127 Myrtle Street, Elkhart, IN 46514, USA.
Millipore: Millipore (UK) Ltd, The Boulevard, Blackmoor Lane, Watford, Herts WD1 8YN, UK. **Millipore Corporation**, 80 Ashby Road, Bedford, MA 01730, USA.
Molecular Probes: Cambridge Bio-Science, 25 Signet Court, Stourbridge Common Business Centre, Swanns Road, Cambridge CB5 8LA, UK. **Molecular Probes Inc**., 4849 Pitchford Avenue, Eugene, Oregon 97402, USA.
MSE: Fisons Scientific Equipment, Bishop Meadow Road, Loughborough, Leics LE11 0RG, UK.
Nalge Co.: Merck Ltd, Broom Road, Poole, Dorset BH12 4NN, UK. **Nalge Co.**, Box 20365, Rochester, NY 14602–0365, USA.
NuCheck Prep, Incorporated, 109 Main Street West, PO Box 172, Elysian, MN 56028, USA.
Nunc: Life Technologies Ltd, PO Box 35, Trident House, Renfrew Road, Paisley PA3 4EF, UK. **Nunc Inc**., 2000 N. Aurora Road, Naperville, IL 60566, USA.
Nycomed: Nycomed (UK) Ltd, Nycomed House, 2111 Coventry Road, Sheldon, Birmingham B26 3EA, UK. **Robbins Scientific Corp**., 814 San Aleso Avenue, Sunnyvale, CA 94086–1411, USA.
Oxoid: Unipath, Ltd, Wade Road, Basingstoke, Hants RG24 0PW, UK. **Unipath, Inc**., 217 Colonnade Road, Nepean, Ontario K2E 7KS, Canada.
Pharmacia/LKB: Kabi Pharmacia Ltd, Pharmacia LKB Biotechnology Division, Davy Avenue, Knowlhill, Milton Keynes, Bucks MK5 8PH, UK. **Pharmacia LKB Biotechnology Inc**., 800 Centennial Avenue, PO Box 1327, Piscataway, NJ 08855–1327, USA.
Pitman-Moore: Pitman-Moore Limited, Breakspear Road South, Harefield, Uxbridge, Middlexes UB9 6LS, UK. **Pitman-Moore, Inc**., 421 East Hawley, Mundelein, IL 60060, USA.
Sartorius: Sartorius Ltd, Longmead Business Park, Blenheim Road, Epsom, Surrey KT19 9QN, UK. **Sartorius Corp. USA**, 140 Wilber Place, Bohemia, Long Island, NY 11716, USA.

Schleicher and Schuell: Anderman and Co. Ltd, Kingston upon Thames, Surrey KT2 5NH, UK. **Schleicher and Schuell, Inc**., 10 Optical Avenue, Keene, NH 03431, USA.
Scientific Programming Enterprises, Haslett, MI 48840, USA.
Serva: Universal Biologicals Ltd, 30 Merton Road, London SW18 1QY, UK. **Serva Fine Biochemicals Inc**., 200 Shames Drive, Westbury, NY 11590, USA.
Sigma: Sigma Chemical Co. Ltd, Fancy Road, Poole, Dorset BH17 7TG, UK. **Sigma Chemical Company**, PO Box 14508, St. Louis, MO 63178, USA.
Sochibo: Sochibo, 3–5 rue Carnot, 92100 Boulogne sur Seine, France.
Sorvall: DuPont (UK) Ltd, Diagnostics and Biotechnology Systems Division, Wedgwood Way, Stevenage, Herts SG1 4QN, UK. **DuPont Co**, Biotechnology Systems, PO Box 80024, Wilmington, DE 19880–0024, USA.
Stephens Instruments, 1251–D Georgetown Road, Lexington, KY 40511, USA.
Supelco: R. B. Radley and Co. (Supelco), Shire Hill, Saffron Walden, Essex CB11 3A3, UK. **Supelco**, Supelco Park, Bellefonte, PA 16823–0048, USA.
Technicon: Technicon Instruments Co. Ltd, Evanshouse, Hamilton Close, Houndmills, Basingstoke, Hants RG21 1BZ, UK. **Technicon Instruments**, 511 Benedict Avenue, Tarrytown, NY 10591, USA.
Union Carbide, Sales Office, Long Beach, CA 92623, USA.
Vector: Vector Laboratories, 16 Wulfric Square, Bretton, Peterborough PE3 8RF, UK. **Vector Laboratories**, 30 Ingold Road, Burlingame, CA 94010, USA.
Vortex: Arnold Horwell Ltd, 73 Maygrove Road, West Hampstead, London NW6 2BP, UK. **Scientific Industries, Inc**., 70 Orville Drive, Airport International Plaza, Bohemia, NY 11716, USA.
Wellcome: Wellcome Diagnostics, Temple Hill, Dartford, Kent DA1 5AH, UK. **Wellmark Diagnostics Ltd**, 650 Woodlawn Road, Guelph, Ontario N1K 1B8, Canada.
Whatman: Whatman LabSales Ltd, St. Leonard's Road, 20/20 Maidstone, Kent ME16 0LS, UK. **Whatman LabSales Inc**., 5285 N.E. Elam Young Parkway, Suite A400, Hillsboro, Oregon 97124, USA.
Wolfson Research Laboratories, Clinical Chemistry Department, Queen Elizabeth Medical Centre, Edgbaston, Birmingham B15 2TH, UK.

Index